Growing Bulbs

A.G.W. Simpson

Kangaroo Press

Cover: Narcissus 'King Alfred'

© A.G.W. Simpson 1985

First published in 1985 by Kangaroo Press
3 Whitehall Road (PO Box 75) Kenthurst 2154
Typeset by T. & H. Bayfield.
Printed in Hong Kong by Colorcraft Ltd

ISBN 0-86417-044-0

Contents

Contents

1 Bulbs for All Occasions

You don't have to be a professional gardener to grow great bulbs. Many of the world's top bulb growers are hobby-gardeners, whose secret of success is to know how, why, what, where and when to treat their bulbs. I would like to share with you the special secrets that I have learned from my long experience world-wide as a professional gardener.

Let us once and for all put the expression 'greenfingers' into proper perspective. There is no such person as a naturally gifted greenfingered gardener. The so-called greenfingered person is one who will spend time with and give devotion to his or her bulbs, thus detecting immediately when things are going wrong or things need to be done. It is that simple.

Would you commit murder to obtain a bulb?

Of course you wouldn't, but it has happened many, many times. Admittedly the bulbs, roughly the size of a table-tennis ball, were worth $2000 or more. But today you can obtain better bulbs at a fraction of that cost. Turn to the section on tulips to learn more about 'tulip madness'.

When is a bulb not a bulb?

When it is a tuber, corm or rhizome. Although for convenience we group bulbs, tubers, corms and rhizomes together collectively as bulbs, they are, nonetheless, distinctively different.

The easiest true bulb to find is an onion. Cut down the centre of the onion and observe that it is made up of scales, which enclose the flowering bud at the centre. Other true bulbs include daffodils, grape hyacinths, lilies, snowdrops and tulips.

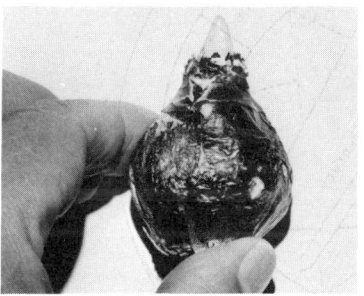

Hyacinth bulb.

Gladiolus corms.

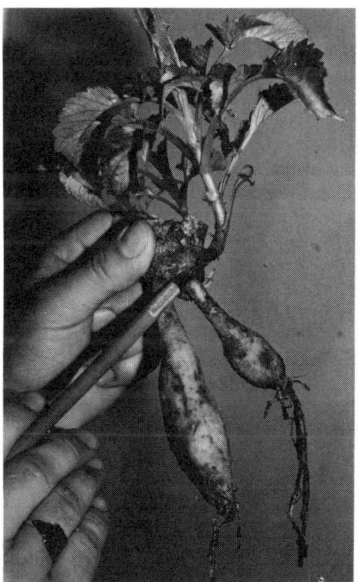

Dahlia tuber. Notice how the shoots appear at the collar *above* the tubers.

Ranunculus tubers. Notice tuber claws.

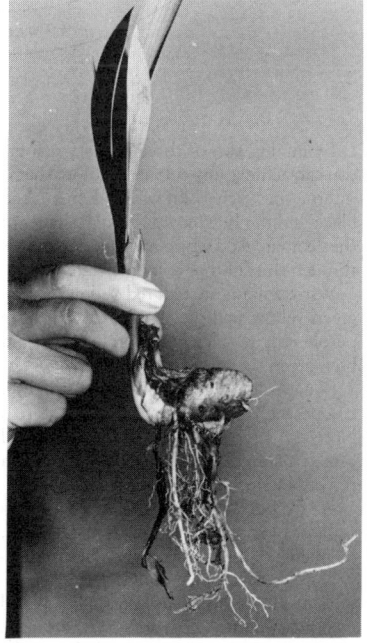

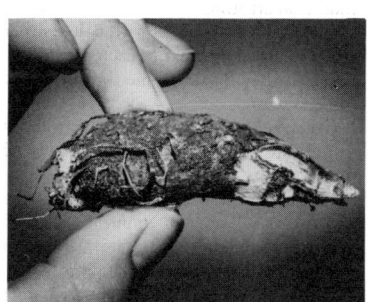

Tuberose bulb.

Iris rhizome.

Canna tuberous root.

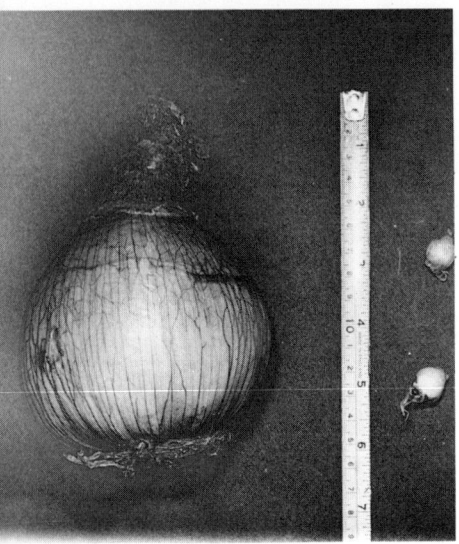

Lilium bulbs come in varying sizes.

Left: Hippeastrum bulb — *Right:* Grape hyacinth bulbs.

Now get a potato. It is a tuber. Examine it and you will notice it has many 'eyes', and just below the eyes are small scars. These 'eyes' are the plant's shoots. Other tubers are dahlias, anemones, begonias (tuberous), cyclamen and gloxinias.

Gladiolus is the most famous corm. Corms are usually flattish. The corm sends out flowers and dies, but not until it has produced new corms above the old, and thousands of pieces of spawn, like grains of rice. Other corms include crocus, freesia, sparaxis and watsonia.

You will probably be familiar with the bearded iris, found throughout Australia. This is a rhizome. The thick, fleshy roots, which look like large bread rolls, are underground stems and contain all the plant characteristics. Examine the root and see where the buds have formed. Other examples of rhizomes are agapanthus and cannas.

Having divided these bulbous root systems into bulbs, tubers, corms and rhizomes, I shall, when convenient to do so, refer to them collectively as bulbs.

Two of the secrets of successful bulb growing are, firstly, always buy top quality bulbs of the best varieties, and, secondly, understand the principles of colour combination.

Colour Combination

There are three primary colours: red, yellow and blue. Mix red and yellow and you get orange. Mix red and blue and you get purple. Mix yellow and blue and you get green. Orange, purple and green are secondary colours. Mix these secondary colours, and you get intermediate, or tertiary, colours such as orange-yellow, blue-green, yellow-green and so on.

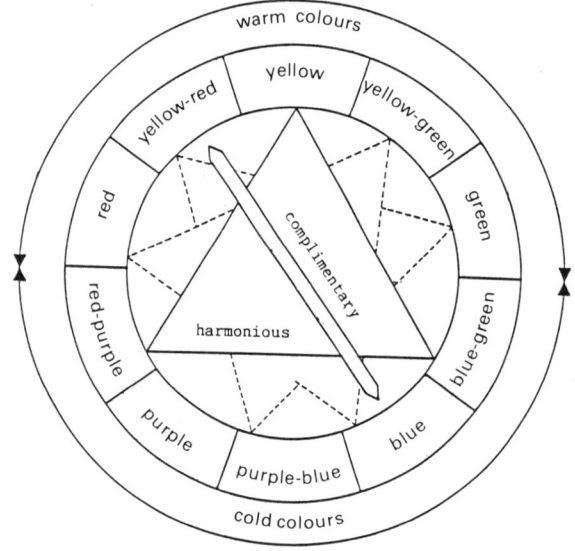

A 'Colour Wheel'

By mixing two or three primary colours you can achieve any colour. A colour that is intense and rich is known as a hue. Add black and the hue moves through shades of that colour. Add white and the hue moves through tints of that colour.

Warm colours are red, red-yellow, yellow, green-yellow and green. Cold colours are purple-red, purple, blue-purple, blue and blue-green.

Some colours advance to catch the eye: colours such as red, orange and orange-red. Note the strength of red traffic lights and orange-painted public vehicles. Too much red or orange will dominate a garden.

A 'Colour Wheel' has been devised to isolate colour combinations (see diagram). Segments of red, red-yellow, yellow, green-yellow, green, blue-green, blue, blue-purple, purple and purple-red, are spaced evenly around the outside of the circle. Inside the circle is an equal-sided triangle,

which can be spun. Those colours where the corners touch will harmonise: for example, blue, yellow and purple-red. Colours that fall opposite will complement each other: for example, red and blue-green. Colours next to each other will complement: red-yellow and yellow. You may be able to buy a 'Colour Wheel' from your local art store.

Can you plant flowers with colours that create disharmony next to each other? Yes, provided they do not come into flower at the same time.

Quick reference guides are given below to indicate a bulb's approximate flowering time, and its suitability for certain areas. More detailed information is given in the main text. Botanical names are used because common names can vary from town to town, state to state, and country to country. However, many common names are given in the main text.

6

Spring Flowering Bulbs

Albuca nelsoni; A. humilis (late)
Allium (most species; late)
Anemone coronaria; A. blanda
Anigozanthos manglesii
Antholyza ringens
Arthropodium cirrhatum
Arum italicum; A. maculatum
Babiana stricta, varieties and cultivars
Billbergia nutans (late)
Boophone disticha
Bomarea caldasii
Brodiaea elegans; B. ida-maia; B. bridgesii; B.
laxa; B. pulchella; B. hyacintha; B. lutea
Bulbinella floribunda; B. hookeri; B. vossii
Calochortus albus; C. amabilis; C. pulchellus;
C. caeruleus
Caloscordum neriniflorum
Camassia quamash (esculenta)) C. leichtlinnii
Chasmanthe aethiopica
Chionodoxa luciliae; C. l. alba; C. l. rosea; C.
l. gigantia; C. sardensis; C. siehei; S.
albescens
Chionoscilla x allenii
Chlidanthus fragrans (late)
Clivia miniata
Colchicum (depending on species)
Colocasia antiquorum; C. a. var. esculenta
(late)
Convallaria majalis and varieties
Corydalis solida; C. emanueli; C.
glaucescens; C. macrocentra
Cotyledon (Umbilicus erectus)
Crocus species and cultivars
Curtonus paniculatus
Cyclamen (depending on species)
Cyrtanthus mackenii cooperi (early)
Dicentra spectabilis; D. formosa; D. eximia
Dierama pulcherrimum; D. pendulum (late)
Dipidax triquetum (triquetra)
Dranunculus vulgaris; D. muscivorus
Eminium albertii
Endymion non-scriptus; E. hispanica
Eranthis hyemalis; E. x tubergenii (early)
Eremurus species (late)
Erythronium species
Eucharis grandiflora; E. candida
Ferraria crispa
Freesia varieties and cultivars
Fritillaria species
Gagea lutea; G. fibrosa
Galanthus species and cultivars
Geranium tuberosum
Gynandriris sisyrynchium
Haemanthus natalensis
Hermodactylus tuberosus (early)
Hesperantha inflexa; H. stanfordiae (early)
Hippeastrum vittatum and cultivars (late)
Hosta species (late to early summer)
Hyacinthella lineata; H. siirtensis; H.
leucophaea
Hyancinthoides non-scripta; H. hispanica (see
Endymion)
Hyacinthus species and cultivars
Incarvillea younghusbandi; I. delavayi
Ipheion uniflora
Iris (Dutch, Dwarf)

Iris (bearded, beardless; late)
Ixia maculata cultivars; I. viridiflora
Ixiolirion tartaricum (montanum)
Lachenalia species and cultivars
Lapeyrousia (Lapeirousia) cruenta
Leontice armenica
Leopoldia cosmosa
Leucocoryne ixioides
Leucojum vernum
Lysichitum americanum
Moraea species
Muscari species and cultivars
Narcissus species and cultivars
Ornithogalum thyrsoides; O. arabicum; O.
umbellatum
Oxalis species and cultivars
Paris polyphylla
Polygonatum multiflorum
Primula fedtschenkoi
Puschkinia scilloides
Ranunculus species and cultivars
Rechsteineria (Sinningia) cardinalis
Romulea selected species
Sanguinaria canadensis (early)
Satyrium species
Sauromatum guttatum
Scilla bifolia; S. peruviana; S. sibirica
Sisyrinchium campestre; S. bermudiana; S.
douglasii
Soldanella pusilla
Sparaxis tricolor cultivars
Spiloxine (Hypoxis) late
Sprekelia formosissima (late)
Sterbergia fischerana S. candida
Strelitzia species
Streptanthera cuprea
Tradescantia virginiana (late)
Trillium species (late or early summer)
Tritonia crocata and cultivars
Tropaeolum species (late to summer)
Tulbaghia species (late to summer)
Tulipa species and cultivars
Veltheimia viridifolia (also early summer)
Wachendorfia thyrsiflora (also early
summer)
Watsonia cultivars
Zantedeschia aethiopica
Zigadenus fremontii

Summer Flowering Bulbs

Achimenes longiflora
Acidanthera bicolor; A. b. murielae (late)
Agapanthus species
Allium species (early)
Alpinia species
Alstroemeria aurantiaca and cultivars
Amaryllis belladonna (late)
Ammocharis coranica
Anigozanthos flavidus; A. pulcherrimus; A.
rufus; A. viridus
Anthurium scherzeranium
Aristea thyrsiflora; A. eckloni; A. macrocarpa
(early)
Begonia tuberosa and cultivars

Belamcanda chinensis
Blandfordia grandiflora; B. nobilis; B. punicea
Bletilla striata (early)
Bomarea caldasii
Bowiea volubilis
Bravoa geminiflora
Brodiaea elegans; B. ida-maia; B. bridgesii
(early)
Brunsvigia josephinae (late)
Bulbine latifolia
Caladium bicolor cultivars
Calachortus species (early)
Calostemma pupurea (late)
Cardiochrinum species
Canna indica and cultivars
Codonopsis convulvulacea
Commellina coelestis
Cooperia pedunculata
Crinum species and cultivars
Crocosmia x crocosmiiflora (late)
Curcuma petiolata
Cypella herbertii
Cyrtanthes angustifolius
Dahlia cultivars
Dicentra species (early)
Dierama pendulum; D. pulcherrium (early)
Dietes (see Moraea)
Disa uniflora
Eremurus stenophyllus; E. olgae; E.
lactiflorus; E. robustus
Eucharis grandiflora; E. candida
(occasionally)
Eucomis species
Freesia cultivars (early)
Galtonia candicans
Gladiolus species and cultivars
Gloriosa rothschildiana; G. superba; G.
simplex (late)
Gloxinia hybrids
Haemanthus katherinae
Hedychium species (late)
Heliconia species
Hemerocallis species
Herbertia platensis
Hippeastrum vittatum and cultivars (early)
Hosta species
Hymenocallis calathina
Hypoxis (Spiloxene) early
Iris (early; bearded; I. tingitana)
Kniphofia species (early)
Lapeyrousia species (early)
Leucojum aestivum
Liatris spicata
Lilium species and cultivars (early-mid-
late)
Liriope muscari and cultivars; L. graminifolia
(late)
Littonia modesta
Manfreda longiflora (early)
Milla biflora
Monstera deliciosa
Musa basjoo; M. rosacea; M. textilis
Neomarica caerulea (early)
Nelumbo lotus
Nymphaea (water lillies)
Nymphoides (water flowers)
Oxalis (early)

Pancratium foetidum P. illyricum; P. maritinum
Phaeomeria magnifica
Polianthes tuberosa (tuberose)
Prochnyanthes viridescens
Pseudobravoa densiflora
Pseudogaltonia clavata
Rhodohypoxis baurii (late)
Rigidella immaculata; R. flammea
Roscoae cautleoides; R. humeana; R. alpina
Sandersonia aurantiaca
Satyrium species (early summer)
Schizostylis coccinea and cultivars (late)
Spathiphyllum species and cultivars
Sprekelia formosissima (early)
Strelitzia species (early)
Tigridia pavonia and cultivars; T. pringlei
Tradescantia virginiana (early)
Tritonia species (early)
Tropaeolum species (early)
Tulbaghia violaceae
Umbilicus erectus
Urceolina urceolata
Vallota speciosa (late)
Wachendorfia thyrsiflora (early)
Watsonia cultivars (early)
Zantedeschia (yellow and pink cultivars)
Zephyranthes grandiflora; Z. candida; Z. citrina

Autumn Flowering Bulbs

Acidantherera bicolor
Amacrinum x
Amaryllis belladonna
Anigozanthos pulcherrimus; A. rufus
Brunsvigia josephinae (early)
Colchicum (autumn species and cultivars)
Crinum moorei (early)
Crocosmia x crocosmiiflora
Crocus sativus; C. longiflorus; C. kotschanus
Curtonus paniculatus
Cybistetes longifolia
Cyclamen (certain species)
Cyrtanthus sanguineus
Dahlia cultivars (early)
Habranthus species
Haemanthus coccinea (early)
Hedychium species (early)
Herbertia platensis (early)
Hymenocallis species (early)
Hypoxis species (early)
Iris unguicularis
Leucojum autumnalis
Liriope muscari and cultivars
Lycoris radiata and cultivars (early)
Montbretia (see Crocosmia)
Nerine species and cultivars
Polianthes (tuberose) (early)
Schizostylis coccinea (early)
Smithiantha zebrina
Sternbergia lutea; S. l. s.sp sicula; S. colchiflora
Streptocarpus cultivars
Tuberose (see Polianthes)

Winter Flowering Bulbs

Anemones (planted late summer)
Anigozanthos humilis; A. manglesii
Clivia miniata (late)
Cyrtanthus mackenii cooperi (late)
Eranthis (species)
Galanthus nivalis (late)
Hermodactylis tuberosa (late)
Kniphofia species (late)
Lachenalia species (late)
Muscari species (late)
Nectaroscordum siculum (late)
Nerine species (early)
Ranunculus (planted late summer)
Tulbghia fragrans
Narcissus, 'Paperwhite'; 'Soleil D'Or'

Popular Bulbs, First Choice

Agapanthus species
Alstroemeria aurantiaca
Amaryllis belladonna
Anigozanthos species and cultivars
Anemone coronaria; 'De Caen'; 'St Brigid'
Babiana cultivars
Begonia tuberhybrida cultivars
Canna indica cultivars
Clivia miniata
Crinum species and cultivars
Cyclamen persicum hybrids
Dahlia cultivars
Eucomis cosmosa
Freesia cultivars
Gladiolus cultivars
Hemerocallis aurantica major cultivars
Hippeastrum vittatum cultivars
Hyacinthus cultivars
Iris bearded; Louisiana; Spuria
Kniphofia uvaria cultivars
Lachenalia species and cultivars
Lilium species and cultivars
Muscari (grape hyacinth)
Narcissus (daffodils) species and cultivars
Nerine species and cultivars
Polianthes (tuberose)
Ranunculus asiaticus cultivars
Sinningia (gloxinia) cultivars
Sparaxis tricolor cultivars
Spekelia formosissima
Strelitzia reginae
Tigridia pavonia
Tritonia crocata
Tulbaghia fragrans
Tulipa species and cultivars
Watsonia cultivars
Zantedeschia species and cultivars

Popular Bulbs, Second Choice

Acidanthera bicolor cultivars
Alpinia species
Arum paleastinum

Chlidanthus fragrans
Cyrtanthus species
Dierama species
Eucharis grandiflora
Eucomis bicolor; E. pole-evansi
Galanthus (snowdrop) species and cultivars
Galtonia candicans
Hedychium species
Hymenocallis species
Iris (all except those mentioned above)
Ixia cultivars
Leucojum (snowflake) species
Liriope muscari
Lycoris africana; L. radiata
Moraea (Dietes) villosa; M. bicolor
Ornithogalum species
Pancratium species
Schizostylis coccinea cultivars
Tulbaghia violaceae
Vallota speciosa
Veltheimia viridifolia and cultivars
Zephyranthes grandiflora; Z. candida; Z. citrina

Popular Bulbs, Third Choice

Achimenes longiflora
Albuca nelsoni; A. humilis
Allium (less vigorous species and cultivars)
Alocasia macrorrhiza
Anthericum liliago
Antholyza ringens
Belamcanda chinensis
Billbergia nutans
Blandfordia grandiflora
Bletilla striata
Brodiaea (Triteleia-Dichelostemma) species and cultivars
Brunsvigia josephinae
Bulbine latifolium
Bulbinella floribunda
Caladium bicolor cultivars
Calochortus species
Calostemma purpureum
Camassia species
Cardiocrinum giganteum
Chasmanthe aethiopica
Chionodoxa species
Colchicum species and cultivars
Colocasia antiquorum and varieties
Commelina coelestis
Crocosmia (Montbretia) x crocosmiiflora
Crocus species and cultivars
Cypella herbertii
Dicentra spectabilis
Endymion (bluebells) species and cultivars
Eranthis species and hybrids
Eremurus species
Erythronium species and cultivars
Fritillaria species and cultivars
Gladiolus species
Gloriosa rothschildiana; G. superba; G. simplex
Habranthus species
Haemanthus species
Hermodactylis tuberosus

Hesperantha species
Hosta species
Ipheion uniflora and cultivars
Lapeyrousia (Lapeirousia) laxa (cruenta)
Leucocoryne ixioides
Liatris spicata
Littonia modesta
Lysichitum americanum
Monstera deliciosa
Musa species
Nelumbo (lotus) nucifera
Neomarica caerulea
Nymphaea (water lilies)
Ophiopogon species and varieties
Oxalis cultivars
Polygonatum multiflorum
Rechsteineria cardinalis
Romulea bulbocodium
Sandersonia aurantiaca
Scilla species and cultivars
Sisyrinchium species
Smithiantha zebrina and cultivars
Spathiphyllum species and cultivars
Sternbergia species and cultivars
Streptanthera species
Streptocarpus cultivars
Tradescantia virginiana
Tricyrtis species
Trillium species
Tropaeolum species
Zingiber officinale

Fritillaria species
Galanthus species and cultivars
Galtonia candicans
Geissorhiza species
Gladiolus species
Gloriosa species
Gyandriris setifolia
Haemanthus species
Heliconia species
Herbertia platensis
Hermodactylis tuberosus
Hippeastrum vittatum new cultivars
Hosta species
Ixiolirion tartaricum
Leucocoryne ixioides
Liatris spicata
Lysichitum species
Nectaroscordum siculum
Neomarica caerulea
Nerine bowdeni
Nomocharis species
Notholirion macrophyllum
Orchids (Australian native)
Polygonatum multiflorum and varieties
Puschkinia scilloides
Rhodohypoxis baurii
Roscoea species
Sanguinaria canadensis
Sandersonia aurantiaca
Soldanella pusilla

Sternbergia species
Tricyrtis stolonifera
Trillium species
Veltheimia cultivars

Bulbs for Pots or Containers

When growing bulbs in pots or containers (such as clean half casks or Versailles tubs), four important points must be remembered. Firstly, considerable attention must be given to the careful watering of the bulbs. Secondly, proper feeding must be adhered to — if necessary, with a suitable fertiliser every year, generally a complete fertiliser with a low nitrogen content. The fertiliser bag usually states if it is suitable for bulbs. Thirdly, repotting each year, if necessary, must be carried out. Fourthly, a suitable compost must be used, for example, John Innes No 1 or any other similar loam, sand and peat compost. Loamless composts can be used provided there is not too much peat in the make-up of the compost, which can dry out too savagely. Sharp sand or grit can be added to open up such a mix. The compost must be damp enough, when planting, to absorb water, as dry compost does not readily absorb water.

Bulbs for Collectors

The bulbs listed below are either not easy to grow, not popular, not readily available, in the forefront of a particular bulb's development, or could become scarce in Australia through lack of interest. Bulb collectors are pathfinders for the rest of us. They find scarce bulbs, create a demand for a particular bulb, or develop societies that encourage interest in unusual bulbs.

Albuca nelsoni; A. humilis
Anigozanthos species and cultivars
Anthurium scherzerianum
Arisaema candidissimum
Arisarum vulgare
Belamcanda chinensis
Biarum davisii
Blandfordia species
Boomarea species
Boophone disticha
Bravoa geminiflora
Bongardia chrysogonum
Brunsvigia josephinae
Caladium species and cultivars
Calochortus species
Camassia species
Colchicum species
Commelina coelestis
Crocus species and cultivars
Crinum cultivars
Cypella herbertii
Dranunculus species
Eremurus species
Erythronium species
Eucharis grandiflora

Pots lined up for planting.

Arrange the bulbs, in this case liliums, by their individual pot; complete with name. The pots can be numbered, but more discreetly than above.

The bulb, a lilium, is placed well into the compost.

The bulb is labelled.

Potting bulbs. Crocking a pot with broken pots.

Potting large spreading bulbs. Prepare a container. Note drainage hole.

Potting large spreading bulbs. Bottom of cask is crocked with broken pots.

Potting large spreading bulbs. Layer of pebbles placed over crocks.

You can grow bulbs in plastic or clay pots, which preferably should be plunged into sand so that they do not dry out too quickly. The ideal combination for appearance and usefulness would appear to be a compost such as John Innes No 1, or similar, and clay pots. New clay pots can be soaked thoroughly in clean water prior to planting. Old clay pots, plastic pots and containers, must be thoroughly scrubbed before being used to grow blubs.

Repotting is usefully carried out in the dormant season, generally just before the bulbs start actively growing, to prevent possible shrivelling brought about by dormant bulbs being in contact with fresh compost. Check text for information on specific bulbs.

Bulbs in pots and containers can be moved into different temperature micro-climates and light-value areas; in other words they can be kept shady when need be, and then moved into the light when their growth pattern dictates, which means that you have more chance of growing difficult bulbs.

The depth to which bulbs should be planted in the pot depends on the individual bulbs. Stem lilies, for example, have to be planted deeply, whereas some other bulbs like to be planted with half the bulb above the soil. Check text for details.

It is important to crock clay pots using broken clay pots or inert rock to allow surplus water to drain out. The crocks are laid *convexly* over the drainage hole so as not to block it up and prevent water escaping. You can use large pebbles to drain casks and other large wooden or clay containers.

Hyacinths can be planted in wooden boxes, and when in flower can be transferred into bowls or pots. This is the only way you will get the startling 'hyacinth effect'; that is, all one colour, or different colours, at the same height and flowering time. It's a guesswork job otherwise. Two layers of daffodils can be planted in one large pot, one layer on top of the other. They should all flower at the same time.

Narcissus such as 'Paperwhite' and 'Soleil d'or', and Roman hyacinths, make useful early bulbs. You can grow hyacinths in willowy, hyacinth-glass jars that have nipped-in waists. The bulb sits in a cup suspended just above the water contained in the glass. The roots of the hyacinth will move down into the jar and appear like coils of white wire. You can buy hyacinth jars in various colours. Keep a constant check to make sure the water has not gone foul. Some people place black wood charcoal in the jar to keep it sweet. You can also grow early narcissus and hyacinths on pebbles,

the bulbs protruding above the water contained in a bowl.

Note well. You or a member of your family may be allergic to certain plants. Be aware of this before growing pot-plants.

Achimenes longiflora
Agapanthus (dwarf species)
Allium (certain cultivars)
Amaryllis hippeastrum
Anthurium scherzerianum
Arthropodium cirrhatum
Babiana cultivars
Begonia tuberhybrida
Billbergia nutans
Blandfordia grandiflora
Bletilla striata
Bulbocodium vernum
Caladium species and cultivars
Canna (smaller variegated-leaved)
Clivia miniata
Crinum x powellii; C. moorei
Crocus species and cultivars
Cyclamen persicum (florists' cultivars); *C.* species
Cyrtanthus species
Freesia cultivars
Gladiolus (small species)
Gloxinia (Sinningia) cultivars
Heamanthus species
Hippeastrum cultivars
Hosta species

Hyacinthus cultivars
Ipheion uniflorum species and cultivars
Iris, bulbous
Ixia cultivars
Lachenalia species and cultivars
Leucojum species and cultivars
Lilium species and cultivars
Liriope muscari
Littonia modesta
Lycoris species
Monstera deliciosa
Moraea species and varieties
Muscari species and cultivars
Narcissus species and cultivars
Neomarica caerulea
Nerine species
Nymphaea (water lilies) in tubs under water
Orchids, native
Oxalis (less vigorous species)
Polianthes (tuberose)
Polygonatum species
Rechsteineria cardinalis
Rhodohypoxis baurii
Romulea species
Sandersonia aurantiaca
Schizostylis coccinea
Scilla species and cultivars
Smithiantha zebrina and cultivars
Sparaxis tricolor and cultivars
Sprekelia formosissima
Strelitzia reginae; S. reginae juncifolia
Streptocarpus cultivars
Tradescantia virginiana
Trillium species
Tulipa species and cultivars
Tulbaghia fragrans
Umbilicus erectus
Vallota speciosa
Veltheimia species and cultivars
Zantedeschia species and cultivars

Subtropical Bulbs

There are many useful, beautiful bulbous plants suitable for subtropical gardens, and the list of species and cultivars given below contains such specimens.

Achimenes longiflora
Acidanthera bicolor
Agapanthus africanus
Alocasia macrorrhiza; A. m. var. rubra; A. m. 'Variegata'
Alpinia speciosa (mutica); A. pupurata
Alstroemeria aurantiaca aurea and cultivars; A. pulchella
Amaryllis belladonna
Anemone coronaria
Anigozanthos species
Anthurium scherzerianum; A. andreanum; A. a. 'Rubrum'
Babiana stricta and cultivars
Begonia tuberhybrida
Billbergia nutans
Blandfordia grandiflora
Brunsvigia josephinae

Caladium colour and cultivars
Canna indica and cultivars
Clivia miniata
Colocasia antiquorum; C. a. var. esculenta
Crinum x powelli; C. moorei; C. bulbispermum
Crocosmia x crocosmiiflora (Tritonia crocosmiiflora)
Curcuma petiolata
Cyrtanthus mackenii, C. m. var. cooperi; C. sanguineus
Dahlia cultivars
Dierama pulcherrima
Dietes (Moraea) bicolor; D. vegeta
Eucharis grandiflora
Eucomis cosmosa
Freesia cultivars
Gladiolus cultivars
Gloxinia cultivars
Habranthus robustus
Haemanthus coccineus; H. multiflorus
Hedychium gardneranum; H. coronarium
Heliconia collinsiana; H. bihai; H. humilis; H. bicolor
Hemerocallis aurantiaca and cultivars; H. fulva
Hippeastrum vittatum and cultivars
Hosta species (late)
Iris, bearded cultivars
Kniphofia uvaria
Lachenalia aloides and cultivars
Leucojum vernum
Lilium longiflorum; L. tigrinium; L. maculatum
Liriope muscari
Littonia modesta
Lycoris africana; L. radiata and cultivars
Montbretia (see Crocosmia)
Moreae spathacea; M. bicolor; M. vegeta
Musa species
Neomarica gracilis
Nerine sarniensis
Ophiopogon japonicus; O. jaburan var. aureus variegatum
Ornithogalum thyrsoides
Phaeomeria magnifica
Polianthes tuberosa
Sinningia (see Gloxinia)
Sparaxis tricolor
Spathiphyllum blandnum; S. patinii and cultivars
Sprekelia formosissima
Strelitzia reginae; S. nicholai
Streptocarpus hybrids
Tigridia pavonia
Tulbaghia violacea
Tritonia crocata
Tuberose (see Polianthes)
Veltheimia viridifolia
Watsonia cultivars
Zantedeschia aethiopica; Z. elliotana
Zephyranthes candida; Z. grandiflora; Z. rosea
Zingiber zerumbet

Principal Bulb Families

Daffodil Family (Amaryllidaceae)
Dahlia–Daisy Family (Asteraceae)

Arum Lily Family (Araceae)
Begonia Family (Begoniaceae)
Blood Lily Family (Haemodoraceae)
Canna Family (Cannaceae)
Ginger Family (Zingiberaceae)
Gloxinia Family (Gesneriaceae)
Iris Family (Iridaceae)
Lily Family (Liliaceae)
Orchid Family (Orchidaceae)
Oxalis Family (Oxalidaceae)
Pineapple Family (Bromeliadaceae)
Poppy Family (Papaveraceae)
Primrose–Cyclamen Family (Primulaceae)
Buttercup-Ranunculus Family (Ranunculaceae)
Strelitzia–Banana Family (Musaceae)
Water Lily Family (Nymphaceae)

Bulbs as 'Accent' Plants

'Accent' plants should be tall enough to demand attention, colourful, slender, vertical, column-shaped, or with a roughly pointed bulb flower cluster. They are used in the garden to grab attention, similar to the artistic placing of a statue. You have possibly seen how this effect is created in botanic gardens. The same eye-catching effect can be achieved in the garden, shadehouse and greenhouse with the subtle use of bulb species and cultivars.

A bulbous 'accent-dramatic' feature can be better explained if one considers the eye-catching effect of suddenly coming upon a group of agapanthus lilies in flower. The dramatic, grouped effect stops you in your tracks, and therefore these agapanthus 'accent' part of the garden. 'Accent' bulbous plants can be most colourful either in leaf or flower or both.

'Accent' bulbs should be used when you wish to draw people's attention to a place. The size and height of the grouping would depend on the size of the garden but the following bulbs should be considered:

Agapanthus orientalis
Allium christophi (albopilosum)
Alpinia speciosa; A. zerumbet
Amaryllis belladonna
Anigozanthos (tall species)
Bletilla striata
Caladium cultivars
Canna indica (tall species and cultivars)
Cardiocrinum species and varieties
Clivia miniata
Colocasia antiquorum species and varieties
Crocosmia x crocosmiiflora
Crinum species and cultivars
Dierama species
Dranuculus vulgaris
Eremurus species
Hedychium species
Hemerocallis species and cultivars
Hippeastrum vittatum cultivars
Hosta species
Iris (tall species and cultivars)
Hymenocallis species and cultivars

Kniphofia uvaria and cultivars
Liatris spicata
Lilium species and cultivars
Lysichitum americanum
Moraea (tall species)
Musa basjoo; M. textilis; Musa rosacea
Ornithogalum caudatum
Zingiber officinale

Bulbs for Damp Places

Some gardens or parts of a garden are cursed or blessed with damp areas. Many bulbs can take dampness for a time, but have to have a drying period. Other bulbs can take constant dampness. Some plants will thrive in moist soils. The bulbs listed below can be planted near or even *in* water:

Arisaema candidissium (damp places)
Aristea thyrsiflora (moist-open-sheltered)
Arum species (rich-moist)
Crinum bulbispermum; C. campanulatum (moist)
Dicentra species (moist)
Dipidax triquetrum (boggy)
Hosta (moist soil only)
Iris kaempferi (occasionally, not permanently boggy)
Iris laevigata (moist-acid)
Iris sibirica (moist-acid)
Iris 'Louisiana' (damp)
Iris 'Spuria' (moisture holding, but drains well)
Iris pseudacoris (wet-marginal areas)
Leucocoryne ixioides (mildly moist)
Leucojum aestivum (damp)
Lysichitum americanum (marshy)
Nelumbo nucifera (boggy-pond)
Nymphaea (water lilies) 450–600 mm (18–24 in) water above bulb
Nymphoides (as above)
Polygonatum multiflorum (moist)
Schizostylis coccinea (moist)
Zantedeschia aethiopica (moist-banks)

Bulbs for Shady Places

A surprising number of bulbs will tolerate a certain amount of shade. However, it must be borne in mind that all green plants need sunlight to manufacture the carbohydrates they need to survive. Remember, too, that light has the ability to wrap itself around shaded objects, which means that light will be absorbed. Look into the shadows and you can see objects clearly. The list below contains mostly bulbs that will tolerate some shade, but need sunlight during part of the day. Those that will tolerate heavy shade are indicated:

Acidanthera bicolor
Agapanthus species
Alocasia macrorrhiza (heavy)
Alpinia species (heavy)
Anthericum liliago
Anthurium scherzerianum (heavy)
Ariseama candidissima

Arisarum species
Aristea species (light)
Caladium species and cultivars
Chasmanthe aethiopica
Clivea species and cultivars
Colchicum species and cultivars (mottled)
Colocasia antiquorum and varieties (heavy)
Convallaria species and varieties
Corydalis species
Crinum moorei
Crocosmia x crocosmiiflora (mottled)
Cyclamen species
Dicentra species (partial)
Dierama species (slight)
Dranunculus vulgaris (partial)
Endymion species and cultivars (mottled)
Eranthis species (mottled)
Erythronium species (mottled)
Eucharis grandiflora
Freesia cultivars (partial in hot areas)
Gloxinia (Sinningia) cultivars
Hedychium species
Hippeastrum species and cultivars (mottled)
Heliconia (heavy)
Hosta species
Hyacinthus species
Iris foetidissima; I. germanica (not cultivars); *I. pseudacorus*
Lachenalia species and cultivars (partial in hot, dry areas)
Lapeyrousia (Lapeirousia) laxa (partial)
Leucojum species
Lilium most species and cultivars (mottled)
Lirope muscari (mottled)
Littonia modesta (partial)
Lycoris species (light)
Lysichiton americanum (partial)
Monstera deliciosa
Muscari species (shade in north)
Narcissus species and cultivars (mottled)
Neomarica caerulea (partial)
Ophiopogon species (heavy)
Polygonatum species
Sandersonia aurantiaca (mottled)
Sanguinaria canadensis
Scilla species and cultivars (partial)
Schizostylis species and hybrids (partial)
Sinningia (see *Gloxinia*)
Smithiantha species and cultivars
Sparaxis tricolor cultivars (partial)
Streptocarpus cultivars
Strelitzia reginae (partial)
Tricyrtis species
Tradescantia virginiana (mottled)
Trillium species
Tropaeolum species
Tulipa species and cultivars
Veltheimia species and cultivars
Zantedeschia aethiopica
Zingiber officinale

Bulbs for Sunny Gardens

Certain bulbs will accept more sunlight than others, and still thrive. Although

there are not as many as one might expect, the list available is still impressive:

Acidanthera bicolor cultivars
Albuca nelsoni; A. humilis
Allium cultivars
Alstroemeria species and cultivars
Amaryllis belladonna
Anemone coronaria; 'De Caen'; 'St Bridgid' strains
Antholyza ringens
Babiana stricta; B. s. rubro cyanea and cultivars
Bilbergia nutans
Brodiaea species and cultivars
Brunsvigia josephinae
Bulbine latifolia
Bulbinella species
Camassia species
Canna species and cultivars
Chlidanthus fragrans
Commelina elegans
Crinum bulbispermum
Cyrtanthus mackenii
Dahlia cultivars
Dierama species
Eremurus species and cultivars
Freesia cultivars (sheltered)
Gladiolus species and cultivars (sheltered)
Hemerocallis species and cultivars
Herbertia platensis
Hesperantha species
Hippeastrum species and cultivars
Hymenocallis species (sheltered)
Hypoxis species
Ipheion uniflora
Iris: bearded; *I. tingitana*
Ixia species and cultivars
Kniphofia species and cultivars
Lachenalia species and hybrids
Lapeyrousia (Lapeirousia) laxa (cruenta)
Liatris spicata
Lilium: certain species (need cool root run, afternoon shade)
Lycoris species
Musa species
Nerine species and cultivars
Ornithogalum species
Oxalis species and cultivars
Polianthes (Tuberose)
Ranunculus asiaticus
Sternbergia species
Strelitzia species
Streptanthera cuprea
Tigridia pavonia and cultivars
Tradescantia virginiana
Vallota speciosa
Watsonia species and cultivars
Veltheimia species and cultivars (partly and sheltered)
Zantedeschia elliotana; Z. rehmannii

Bulbs for Hard Frost Areas

The collection of bulbs given below is suggested for those areas which experience hard frosts. However, it should be borne in mind that summer bulbs such as dahlias can be planted after the risk of frost is over, and lifted before severe frosts begin.

The range of bulbs that can be grown in hard frost areas can be increased by building a bulb frame, which is a four-wall box-foundation of sleepers or bricks or aluminium of any height up to about hip level. The long sides can be 1.2 m (48 in) apart, and as long as is practical.

Lay small mesh wire on the floor-base of the frame to prevent small animals burrowing up through the compost and eating the bulbs. The walls of the frame should be higher than the bulb's flowering height to allow for air circulation and the placing of lights over the frame in inclement weather.

The bottom layer should be 300 mm (12 in) of excellent drainage material such as 16–30 mm (¾–1¼ in) diameter broken rock. This is topped by 175 mm (7 in) of peat-compost, or spent-mushroom compost, or leaf mould compost. This is then topped by 75 mm (3 in) of the compost in which you intend to grow the bulbs — usually a loam, sand and peat mixture with extra grit added to facilitate drainage.

The bulb frame should not be higher than hip level, although some people prefer a higher bulb frame as they find maintenance and viewing of small bulbs more convenient at that height. Generally, you would grow small bulbs in a bulb frame for the practical reason that taller bulbs would stick out of the frame too far. You would not grow small, invasive bulbs in a bulb frame.

An interesting point to note is that many frost-tender bulbs suffer from lack of drainage around the roots, particularly in summer, rather than from frost. If this sounds like a contradiction in terms bear in mind that many bulbs are hard-baked during the summer, which is essential for their survival.

The bulbs can be surrounded by silver, sharp sand to encourage rooting. The frame can then be top-dressed with small broken metal (stone) to facilitate drainage around the neck of the bulbs and set off the frame attractively. The bulbs should be labelled with indelible inked labels or aluminium labels, so that you know which plant is which.

You can fertilise the bulbs in a bulb frame with liquid fertiliser with low nitrogen content. If a bulb fertiliser is not available, use low nitrogen tomato fertiliser.

Many growers place the bulbs in lattice pots (similar to milk crates) filled with compost, of various sizes according to the bulb's size and spread. These can be lifted easily and the bulbs separated with little effort.

The frames can be covered in winter with lights, similar to window frames. On particularly savage nights the lights can in turn be covered by carpets or bracken or peat to keep out the frost. Close observation is necessary to work out whether to remove the lights or replace them should the weather change. The lights should be removed in spring to allow the bulbs to grow naturally, and the bulbs kept watered as need be.

Allium giganteum (pink-violet); *A. christophi (albopilosum)* (white); *A. caeruleum* (lilac-blue); *A. moly* (mimosa yellow); *A. oreophilum* (deep pink)

Amaryllis belladonna (white-pink; sunny wall, mulch in winter)

Chionodoxa lucillae (deep, sky blue); *C. l.* var. *giganteum* (blue); *C. sardensis* (wisteria-violet)

Colchicum autumnale (pale pink)

Crocus angustifolius (susianus) (canary yellow: red-purple reverse); *C. biflorus* (lavender); *C. chrysanthus* and cultivars (white, creamy yellow, canary yellow, mimosa yellow, lilac-reddish purple flush); *C. kotschyanus* s.sp. *cappadocious* (pink); *C. sieberi* (pale lavender); *C. speciosus* (white, pink, pale lilac, lilac-pink); *C. tommasinianus* (purple flower, white tube) *C. vernus* cultivars (white, mimosa yellow, lavender- lilac, royal purple)

Cyclamen cilicium (white); *C. purpurascens (C. europaeum)* pink; *C. coum* (pink); *C. hederifolium (C.neapolitanum)* pink or white.

Endymion non-scripta; *E. hispanica* and cultivars (pink or blue)

Eranthis hyemalis (mimosa yellow)

Erythronium revolutum (light rose pink or creamy white-pink); *E. americanum* (mimosa yellow); *E. a.* var. *castaneum* (brownish orange); *E. californicum* (pale yellow); *E. dens-canis* (pale pink); *E. grandiflorum* (bright yellow); *E. hendersonii* (pink-lilac); *E. oregonum* (pale yellow); *E. tuolumnense* (canary yellow)

Eucomis bicolor (white—needs cover in hard frosts)

Fritallaria alburyana (deep, blotched rose pink); *F. aurea* (mimosa yellow reddish purple spots); *F. crassifolia* (translucent white); *F. imperialis* (mimosa, light orange, sunny wall); *F. meleagris* (white-purple-reddish)

Galanthus nivale and cultivars (white)

Hermodactylis tuberosus (green and purple-black)

Hyacinthus orientalis (blue); *H.* cultivars (blue-pink-yellow)

Ipheion uniflorum (white, pale blue)

Iris (bearded-various colours-mulch bulbs); *I. danfordiae* (yellow); *I.* 'Dutch' hybrids (various blues and yellows; cover tops in winter); *I. kaempferi* (purple-blue-pink-white); *I. latifolia* (lilac-blue); *I. reticulata* (light blue to purple-blue; *I. tectorum* (lavender-blue).

Ixiolirion tartaricum (blue-violet-purple; mulch bulbs)

Leucojum vernum (white)

Lilium species (various colours; apply light, non-packing mulch material)

Muscari species (blue, white)

Narcissus species and cultivars (yellow, white, pinkish)

Polygonatum multiflorum (white)

Scilla sibirica (pale to deep sky blue)

Sternbergia lutea (yellow; mulch)

Tulipa species (various colours)

Bulbs for Beginners

These bulbs are hardy in most parts of temperate Australia and are easy enough to grow. Although many are extremely popular, they are still beautiful. Species are the original plants. Varieties are wild hybrids, found in nature and different from the parent plant. Cultivars are varieties cultivated by humans.

Acidanthera bicolor cultivars
Agapanthus species and cultivars
Anemone coronaria ('St Caen'; 'St Brigid')
Alstroemeria aurantiaca cultivars
Aristea ecklonii
Arthropodium cirrhatum
Babiana stricta cultivars
Bletilla striata
Billbergia nutans
Brodiaea species and cultivars
Canna indica cultivars
Chasmanthe aethiopica cultivars
Chlidanthus fragrans
Clivia miniata
Crinum species and cultivars
Crocosmia x crosmiiflora cultivars
Dahlia cultivars
Daffodil (see *Narcissus*)
Dierama species
Eucomis cosmosa
Freesia cultivars
Galtonia candicans
Gladiolus cultivars
Hemerocallis species and cultivars
Hippeastrum vittatum cultivars
Hymenocallis species
Iris, bearded cultivars
Kniphofia uvaria cultivars
Lachenalia species and cultivars
Leucojum vernum
Liriope muscari
Moraea villosa; *M. bicolor*
Narcissus (Daffodil) cultivars
Ranunculus asiaticus cultivars
Tigridia species and cultivars
Tulbaghia fragrans
Zantedeschia yellow and pink species and cultivars
Zephyranthes species and cultivars

Bulbs for Children

I have found in a lifetime's gardening that children are attracted to small, miniature-

13

flowering bulbs and colourful, small bulbs. However, it is reported that certain bulbs, can cause problems. Bulbs such as *Aconitum napellus* (Monkshood); *Alocasia macrorrhiza* (Spoon Lily-Cunjevoi); *Convallaria majalis* (Lily of the Valley); *Gloriosa species* (Glory lily); Potatoes (green parts); *Zantedeschia aethiopica* (Calla-Arum lily); have resulted in severe poisoning, possibly skin irritation when touched; sometimes death, when eaten. For more information concerning problem plants contact your local Poisons Information Centre (see front of phone book); Dept of Health; Dept Of Agriculture. The magazine *Choice*, April 1983, has an informative article on poisonous plants. (Check your library). Get your bulbs identified correctly. Keep bulbs, pot-plants, indoor pot-plants, and cut-flowers away from small or crawling children.

Miniature and Small Bulbs

Chionodoxa luciliae; *Crocus* cultivars; *Cyclamen* species; *Dahlia*, dwarf-bedding; *Galanthus*; *Iris*, dwarf forms; *Lachenalia*; *Leucojum vernum*; *Muscari*; *Tulipa*.

Bulbs for Soils Rich in Nutrients and Moisture

The selection of bulbs given below contains those species, varieties and cultivars that can be gross feeders, but which also like a soil that is well drained, but has the capacity to hold moisture — in other words, does not dry out rapidly. These soils are known as *humus* soils, and are not to be confused with wet, clay soils.

Achimenes longiflora
Agapanthus species and cultivars
Allium species and cultivars
Alocasia macrorrhiza
Anemone coronaria and cultivars
Aristea thyrsiflora
Arum italicum; *A. maculatum*
Begonia tuberhybrida
Bletilla striata
Bulbinella latifolium
Caladium species and cultivars
Canna species and cultivars
Cardiocrinum species
Clivia miniata and cultivars
Convallaria species and varieties
Crinum bulbispermum
Crocus species and cultivars (lighter soil)
Dahlia cultivars
Dierama species
Eucomis species
Fritillaria meleagris; *F. imperialis*
Galanthus species and varieties
Gloxinia (Sinningia) cultivars
Haemanthus species
Hedychium species
Hemerocallis species and cultivars

Hippeastrum vittatum and cultivars
Hosta species
Hyacinthus species and cultivars
Iris sibirica; 'Pacific Coast Hybrids'; 'Louisiana Hybrids'
Kniphofia species and cultivars
Lilium species and cultivars (not rich in manure)
Moraea species
Narcissus species and cultivars
Neomarica caerulea
Nerine species and cultivars
Nomocharis species
Polianthes tuberosus
Polygonatum multiflorum
Ranunculus species and cultivars
Rechsteineria cardinalis
Roscoea species
Sanguinaria canadensis
Schizostylis coccinea and cultivars
Smithiantha zebrina and cultivars
Streptocarpus cultivars
Strelitzia species
Tigridia pavonia and cultivars
Trillium species
Veltheimia viridifolia and cultivars
Watsonia cultivars
Zantedeschia species and cultivars
Zingiber officinale

Bulbs that Withstand Drought Well

The bulbs given below are those that withstand limited drought conditions well. They would, in their native habitat, expect to suffer regularly from drought as part of their routine existence. Alternatively, some species need a hard baking by the strong, summer sun to mature the bulbs for the following flowering season. Most of them are South African bulbs, but others emanate from various parts of the world:

Anigozanthos species
Arthropodium cirrhatum
Babiana stricta varieties and cultivars
Bilbergia nutans
Brodiaea laxa; *B. lutea*
Bulbine latifolium
Crinum x powelli
Cyrtanthus species
Eranthus cilicicus
Gynandriris sisyrynchium
Gymnospermium alberti
Ipheion uniflorum and cultivars
Ixiolirion tartaricum
Lachenalia species and cultivars
Liriope species
Muscari commutatum
Ophiopogon japonicus
Ornithogalum thyrsoides; *O. arabicum*
Romulea bulbocodium
Sparaxis tricolor cultivars
Sternbergia fischerana
Tulbaghia violaceae

Bulbs that Make Useful Cut-Flowers

Note well. You or a member of your family may be allergic to certain cut-flowers. Be aware of this before using cut-flowers in the house, or other like places.

Acidanthera bicolor cultivars
Agapanthus species and cultivars (flowers and seedheads)
Allium moly
Alstroemeria aurantiaca cultivars
Amaryllis belladonna
Anemone cultivars
Anigozanthos species and cultivars
Anthurium scherzerianum
Begonia tuberhybrida
Billbergia nutans
Babiana varieties and cultivars
Brodiaea species
Chasmanthe species and cultivars
Chlidanthus fragrans
Clivia miniata
Convallaria species and varieties
Crocosmia x crocosmiiflora
Crinum species and cultivars
Cyclamen species
Cyrtanthus species
Dahlia hybrids
Dicentra species
Eucharis species
Eucomis species
Eranthis species
Eremurus species
Freesia cultivars
Galanthus species and varieties
Galtonia candicans
Gladiolus species and cultivars
Haemanthus coccineus
Hermodactylis tuberosus
Hippeastrum species and cultivars
Hyacinthus species and cultivars
Hymenocallis species
Incarvillea species
Iris kaempferi; *I. stylosa*; *I tingitana*; bearded cultivars; Dutch; English; Spanish
Ixia species and cultivars
Ixiolirion tartaricum
Kniphofia species and cultivars
Lachenalia species and cultivars
Leucocoryne species and varieties
Leucojum species
Liatris spicata
Lilium species and cultivars
Liriope muscari
Lycoris species
Morae villosa; *M. bicolor*
Muscari species
Narcissus species and cultivars
Nymphaea cultivars (water lilies)
Nerine species
Ornithogalum species
Polianthes tuberosus (tuberose)
Polygonatum multiflorum
Ranunculus asiaticus cultivars
Schizostylis species and cultivars

Scilla species
Sparaxis tricolor hybrids
Sprekelia formosissima
Strelitzia reginae; S. r. juncea
Tritonia crocata and cultivars
Tulbaghia fragrans
Tulipa species and cultivars
Vallota speciosa
Veltheimia species and cultivars
Watsonia cultivars
Zantedeschia species and cultivars

Bulbs for the Greenhouse

There are bulbs that are too sensitive to be grown outside in the garden or even in a cold greenhouse in certain parts of Australia. Therefore, to grow these delicate bulbs in temperate Australia you will need a heated greenhouse, one that can maintain a temperature of 13–21°C (55–70°F) during the winter. The greenhouse should also be able to be cooled down during summer. There are many systems available.

A cold greenhouse, one that is not heated, can be used to increase your stock of hard-to-grow bulbs in those areas where frost can be a problem.

The greenhouse bulbs listed below generally require 13–21°C (55–70°F) winter temperature.

Achimenes species and cultivars
Anthurium andreanum; A. scherzerianum
Begonia (tuberous and rhizomatous species and cultivars)
Caladium species and cultivars
Cyclamen persicum (florists' cultivars)
Disa uniflora 7.2–10°C (45–50°F)
Gloxinia cultivars
Heliconia species
Herbertia platensis
Rechsteineria cardinalis
Sinningia (*Gloxinia*)
Smithiantha species and cultivars
Spathiphyllum species and cultivars
Streptocarpus species and cultivars

Bulbs for the Seaside

There are few bulbs that will stand up to the rigours of being close to the sea front. Like most desirable shrubs and flowers, they have to have protection before they can colonise an area. Some bulbs, however, do grow in close proximity to the sea: bulbs such as *Pancratium illyricum, Urginea maritima* and *Cyclamen maritimum*. A few more will grow reasonably close to the sea, such as *Agapanthus orientalis, Chlidanthus fragrans* and *Kniphofia uvaria*. A paucity of bulbs indeed! However, once a suitable barrier of shrubs is established between the sea front and the bulbs, more desirable, hardy bulbs can be planted. A concise book on seaside planting should be consulted on the type of protective plants needed.

Bulbs for the Rock- Garden

Many bulbs are suitable for rock gardens: bulbs that are reasonably dwarf; non-invasive; able to grow in a hungry soil (a rich soil would lead to a rankness in plant growth); thrive in a free-draining soil; and be attractive in either flower or foliage or preferably both. Taller bulb species and cultivars would look out of place in a small rock garden, because their size would dominate. However, a group of taller bulbs can be planted in a larger rock garden to establish a strong 'accent' point. Taller bulbs such as *Agapanthus* species, *Nerine flexuosa, Lilium martagon,* and some of the taller species tulips look striking against a dark tree background or dark wall. The rock garden should be in the open, but sheltered, and any bulb that needs shading can be shaded by the rocks, or rock garden trees and shrubs. Conversely, those bulbs that need baking in the summer can be placed to receive full sun.

There are bulbs which people use as rock garden plants, but can be semi-invasive; bulbs such as the more desirable *Allium* (onion) species and cultivars, *Babiana* species and cultivars; *Convallaria* (lily of the valley) species and varieties; *Leucojum* (snowflake) species; *Muscari* (grape hyacinth) species and cultivars; *Zephyranthes* (storm lily) species. If these form part of your rock garden scheme, plant so that they can be controlled easily, perhaps towards the front of the rock garden.

You can spot-treat problem bulbs using the herbicide based on glyphosate, available everywhere, which kills those plants it touches, but doesn't poison the soil. This gives you more chance of containing plants that have become invasive.

Agapanthus species ('accent' plants)

Allium caeruleum; A. flavum; A. moly (place for easy control)
Amana (Tulipa) edulis
Amaryllis belladonna ('accent' plant)
Anemone coronaria cultivars
Bulbocodium vernium; B. veriscolor
Babiana species and cultivars (place for easy control)
Chlidanthus fragrans
Chionodoxa species and varieties
Chionoscilla x allenii
Colchicum species and cultivars
Convallaria majalis varieties (place for easy control)
Crocus species
Cyclamen species
Endymion non-scripta; E. hispanica; cultivars
Eranthis hyemalis; E. cilicicus; varieties and cultivars
Erythronium species and cultivars

Eucomis cosmosa
Fritillaria meleagris
Galanthus species and cultivars
Haemanthus species
Hyacinthella species
Iris danfordiae; I. histrio; I histrioides; I. reticulata; varieties and cultivars
Iris unguicularis
Lachenalia species and cultivars
Leucojum species (place for easy control)
Lilium martagon
Muscari azureum (place for easy control)
Narcissus dwarf species and cultivars
Nerine flexuosa; N. curvifolia var. *fothergillii major; N. filifolia*
Puschkinia scilloides
Scilla bifolia; S. sibirica varieties
Sprekelia formosissima
Sternbergia species
Tulipa; drawf species and cultivars
Vallota speciosa
Zephyranthes grandiflora; Z. candida; Z. citrina (place for easy control)

Bulbs for Garden and Path Edges

Edging bulbs are those used for separating lawns from paths or gardens from paths or car drives, and so on. They can be left in the same site for a number of years and, therefore, the ground should be prepared well and fertilised with a bulb fertiliser prior to planting. If a bulb fertiliser is not available, use a low nitrogen, general purpose fertiliser, such as one used for growing tomatoes. You may wish to remove the bulbs immediately after they have flowered and replace them with seasonal, flowering plants. You may also wish to store the bulbs until they are needed for planting in the correct season. It is generally recommended that the bulb foliage be allowed to die back naturally after flowering to force the goodness from the leaves back into the bulbs. This would be impractical if you replace these bulbs with other plants in the border. Dig out the bulbs with some soil attached to the roots, and 'heel' then in at another part of the garden for the foliage to die down naturally.

Anemone species and cultivars
Chionodoxa species, varieties and cultivars
Convallaria species
Dahlia (dwarf bedding)
Freesia cultivars
Gladiolus cultivars
Hyacinthus cultivars
Lachenalia species and cultivars
Muscari azureum; M. botryoides; M. armeniacum and cultivars
Narcissus (daffodils) cultivars
Nerine species and cultivars
Ranunculus asiaticus cultivars
Tulipa species and cultivars

Bulbs to Grow in Grass

One of the most spectacular sights is a mass of bulbs growing in grass, and many gardeners naturalise bulbs this way. Those of you with kikuyu or buffalo grass will not have much success in naturalising bulbs. Only the so-called winter grasses *Agrostis* (browntop) species, *Festuca* (fescue) species; *Lolium* (ryegrass) species and *Poa* (meadow grass) species are suitable for growing bulbs. Restrict the selection of bulbs to those that do not cause problems; use bulbs such as *Narcissus* (daffodils) species; *Lilium* species and *Tulipa sylvestris*. The bulbs must be left after flowering for the foliage to die down and the goodness to drain back into the bulbs. This is imperative for the bulbs' health.

Scatter narcissus and tulip bulbs by throwing them lightly across the grass area and plant them where they fall; liliums are placed in groups. Using a sharp spade, cut out a turf leaving one side still attached. Dig the area under the cut turf, incorporating sand and peat if necessary to lighten the soil. Plant the bulbs so that the top of the bulb is 50 mm (2 in) below the surface. Roll the turf lightly back into place. You can spread a dressing of low-nitrogen complete fertiliser over the area.

Bulbs for Massed Bedding

Some bulbs look better when massed in beds, and obvious examples are the beds of cannas and agapanthuses we see throughout temperate Australia, particularly in our public parks. In the cooler areas close to Sydney, travellers marvel at the beds of tulips in due season. Tulips can be displayed over taller annual or perennial herbaceous plants such as wallflowers, forget-me-nots, polyanthus and the like.

There are gardeners who like to use lower under-storey plants in combination with bulbs. Examples of these lower under-storey plants are sweet alyssum, ageratum, dark or light blue lobelia, kingfisher daisy (*Felicia* sp.), candytuft (*Iberis* sp.), pansies and violas, dwarf *Phlox drummondii*, English daisies (*Bellis perennis*), and more. The problem with this approach is that weeds are hard to remove, and considerable hand-weeding is needed.

Anemones, narcissuses (daffodils), ranunculuses, amaryllises, agapanthuses, clivias and liliums are also best massed in beds of their own. Although few suburban gardens have much room for the massed bedding of bulbs, even small areas may benefit from such eye-catching treatment.

The area where the bulbs are to be planted must be well dug and all weeds removed. This is followed by a bulb fertiliser dressing, or a complete fertiliser dressing with a low nitrogen content. A list of massed-bedding bulbs follows:

Anemone species and hybrid strains
Acidanthera bicolor cultivars
Alstroemeria species and cultivars
Amaryllis belladonna
Babiana species and cultivars
Begonia tuberhybrida
Bletilla striata
Clivia species and cultivars
Commelina coelestis
Crocosmia x crocosmiiflora
Crocus species and cultivars
Dahlia cultivars
Endymion species and cultivars
Freesia cultivars
Galanthus species, varieties and cultivars
Galtonia candicans
Gladiolus species and cultivars
Hyacinthus cultivars
Hemerocallis species and cultivars
Hymenocallis species and cultivars
Iris species and cultivars
Leucojum species
Lilium species and cultivars
Muscari species and cultivars
Oxalis (non-vigorous species and cultivars)
Rununculus asiaticus cultivars
Scilla species and cultivars
Trillium species
Tritonia species and cultivars
Tulbaghia species
Tulipa species and cultivars
Watsonia cultivars

Bulbs in the Mixed Border

A mixed border is one which contains herbaceous annual and perennial flowers, bulbs and shrubs, which are arranged so that one flower form complements the others. It is also an economical way of having herbaceous flowers, bulbs and shrubs in a confined space.

By referring to our bulb seasonal flowering list, we find that different bulbs will flower at various times of the year. We also know, from studying books on herbaceous annuals, perennials and shrubs, when they will come into flower. The art then is to combine the various forms to have flowers in bloom from herbaceous plants, bulbs and shrubs, over as long a flowering period as possible.

It is in the mixed border that we can use 'Accent-dramatic' bulbs to great effect: bulbs such as *Agapanthus* species; *Amaryllis belladonna*; *Anemone* species and cultivars; *Canna* species and cultivars; *Clivia* species and cultivars; *Galtonia candicans*; *Hemerocallis* species and cultivars; *Iris* species and cultivars; *Kniphofia* species and cultivars; *Liatris spicata*; *Lilium* species and cultivars, *Nerine* species and cultivars, *Muscari* species and cultivars; *Narcissus* species and cultivars; *Ranunculus* species and cultivars; and so on.

Avoid using invasive bulbs in a mixed border as they would soon dominate the area. Bulbs such as species and cultivars of *Allium*, *Sparaxis*, *B abiana*, *Chasmanthe*, *Watsonia*, *Homeria*, *Ixia*, *Oxalis*, *Moraea* (*Dietes*), *Romulea*, *Tritonia* and the like should not be used.

By checking the seasonal bulb flowering list, we find that even in winter there are bulbs available, which will liven up the mixed border when other flowers are scarce. Imagine the effects you can create with the host of spring, summer and autumn flowers. Your imagination is your only barrier.

You can also use 'extravagant' bulbs in the mixed border. In summer, bulbs such as *Begonia tuberhybrida*, *Caladium* species and cultivars, *Gloxinia* cultivars and *Streptocarpus* cultivars can be grown in a shady spot, in conditions they would expect to find if they were confined to a greenhouse.

Traditional spring flowering bulbs such as *Narcissus* (daffodil) species and cultivars, *Hyacinthus* cultivars, and *Tulipa* species and cultivars, can be used to great effect in the mixed border.

Bulb Heights

To attempt to classify bulb heights exactly would be difficult, if not impossible, owing to soil differences, local micro-climates and plant variations within the species. However, gardeners like to associate average heights when considering bulb planting. They know, for example, that an agapanthus is a tall bulb, a daffodil is a medium bulb, and a snowdrop is a dwarf bulb. I have therefore classified the bulbs listed below into three convenient heights. 'Tall' bulbs are those above chest height; 'Medium' are those up to chest height; 'Dwarf' are those around ankle height (based on the height of an average person). (d.sp. = dwarf species; t.sp. = tall species; cvs = cultivars or cultivated varieties.)

Tall

Agapanthus	*Colocasia*
Alocasia	*Crinum*
Alpinia	*Dahlia* (t.cvs)
Alstroemeria	*Incarvillea*
Anigozanthos (t.sp.)	*Iris* (t.sp.)
Bomarea	*I. tingitana*
Bowiea	*I. unguicularis*
Bulbinella	*Tradescantia*
Canna	*Tropaeolum*
Cardiocrinum	*Zantedeschia*
Codonopsis	*Zingiber*

Medium

Acidantherea	*Anigozanthos*
Albuca	*Anthurium*
Amaryllis	*Anemone*

Aristea
Arum
Babiana
Begonia
Billbergia
Belamcanda
Brodiaea
Brunsvigia
Caladium
Calochortus
Caloscordum
Calostemma
Cammasia
Chlidanthus
Codonopsis
Clivia
Crocosmia
Cooperia
Cypella
Cyrtanthes
Dahlia
Dicentra
Dipidax
Dranunculus

Eminium
Endymion
Eucharis
Eucomis
Ferrarria
Freesia
Gagea
Galtonia
Geissorhiza
Geranium
Gladiolus
Habranthus
Haemanthus (t.sp.)
Hemerocallis
Hermodactylis
Hippeastrum
Hosta
Hymenocallis
Hypoxis
Incarvillea
Ipheion
Iris
Ixia
Ixiolirion

Kniphofia
Lapeyrousia
Leontice
Leopoldia
Leucojum (t.sp.)
Lilium
Liriope
Lycoris
Milla
Moraea
Narcissus (t.sp. and cvs)
Nectaroscordum
Nerine
Nomocharis
Notholirion
Nothoscordum
Ornithogalum (t.sp.)
Pancratium
Polianthes

Dwarf

Achimenes
Amana
Antholyza

Polygonatum
Ranunculus
Rechsteineria
Rhodohypoxis
Romulea
Roscoae
Sanguinaria
Schizostylis
Sisyrhynchium
Sparaxis
Sprekelia
Streptanthera
Tigridia
Trileleia
Tritonia
Tulbaghia
Tulipa
Umbilicus
Zephyranthes

Bellevalia
Biarium
Bongardia

Brimeura
Bulbocodium
Chionodoxa
Convallaria
Corydalis
Cotyledon
Crocus
Cyclamen
Dahlia (d.sp.)
Eranthis
Erythronium
Fritillaria (d.sp.)
Galanthus
Gagea
Gladiolus (d.sp.)
Gymnospermium
Hyacinthella
Hyacinthus
Haemanthus (d.sp.)
Hesperantha
Ipheion
Iris (d.sp.)
Lachenalia

Lapeyrousia
Leontice
Leucocoryne
Leucojum (d.sp.)
Lloydia
Merendera
Muscari
Ornithogalum (d.sp.)
Oxalis
Puschkinia
Rhodohypoxis
Romulea (d.sp.)
Roscoea
Scilla (d.sp.)
Sparaxis
Streptanthera
Streptocarpus
Sternbergia
Tecophliea
Umbilicus
Urceolina
Zephyranthes
Zigadenus

---

2 Planting Bulbs

Obtaining Bulbs

There are many species and hybrids of bulbs available. However, certain bulbs are difficult to obtain, while others are just not available for Australian and New Zealand gardeners. Bulb growers and importers are introducing a far greater variety of new species and hybrids — one has only to compare what is available today with what was available just a few years ago.

The obvious places to obtain bulb supplies are from bulb growers, bulb importers and seed firms. Check the local and interstate yellow pages in the telephone book. The latter may be found in your local library or post office. Gardening magazines such as *Your Garden* are also informative sources of bulb availability. Your local Parks and Garden Department is another source. Perhaps you have a gardening section in your local paper. Does your local radio run a listeners' gardening programme? A botanical garden advisory service is an obvious source of information. There may be a state Department of Agriculture Home Gardens Advisory Service. All these are potential sources of bulb information. *Do not illegally import or bring bulbs into Australia as you will be contravening the quarantine laws. These laws are very strict as are the penalties for breaking them. State laws should be complied with as well.*

Soil Testing

You may consider applying a dressing of garden lime before planting certain bulbs that like a neutral or limy soil. *Never do this*, at least, not until you have tested the soil to see if it does need lime, as excessive lime could cause problems by locking up essential plant elements such as iron.

Scientists have devised a system for testing the lime content in soil, which is called the 'pH Test'. A scale from pH0 to pH14 has been devised where pH7 is deemed as neutral. From pH7 down to pH0, the soil is becoming progressively more acid (sour). From pH7 up to pH14, the soil is becoming progressively more alkaline (limy).

Most plants thrive well in a soil between pH6.5 and pH7.5. Some bulbs will only thrive in an acid soil — certain liliums for example. Others, such as tulips, will thrive in a alkaline soil.

You can buy simple soil-testing kits from your supplier for testing your soil to see if it is acid or neutral or alkaline. If the soil you are testing is excessively acid or excessively alkaline, then check and double-check your tests, as such readings are not necessarily normal. Take soil samples from all over the area where you are going to plant the bulbs, from the bulb root area, and not from the soil surface.

It is easy to make an acid soil neutral or alkaline, but do you want that? It is nearly impossible to change an alkaline soil into an acid soil. So think carefully before you change the pH values of your soil.

Gypsum is a compound, calcium sulphate, neutral in its soil reaction. However, it does supply calcium, which is the main ingredient of lime (calcium hydroxide or calcium oxide). It also supplies sulphate, which is acid (sulphuric acid). Many gardeners use gypsum to supply their soil calcium needs.

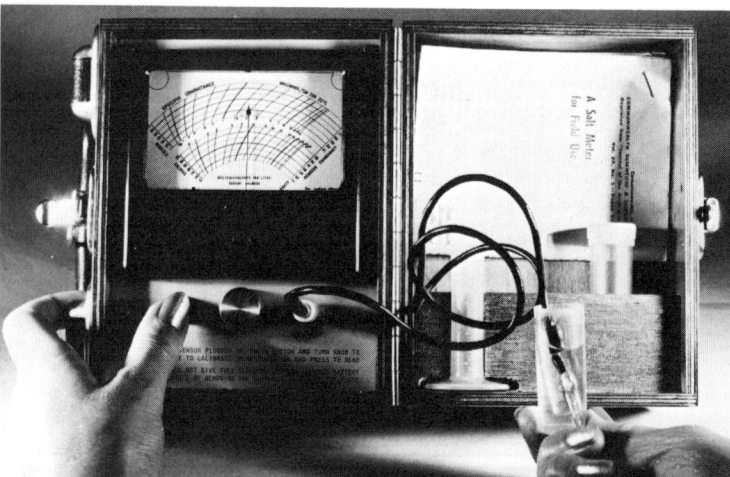

Soil Testing. 'Soil-Salt Content' testing kit.

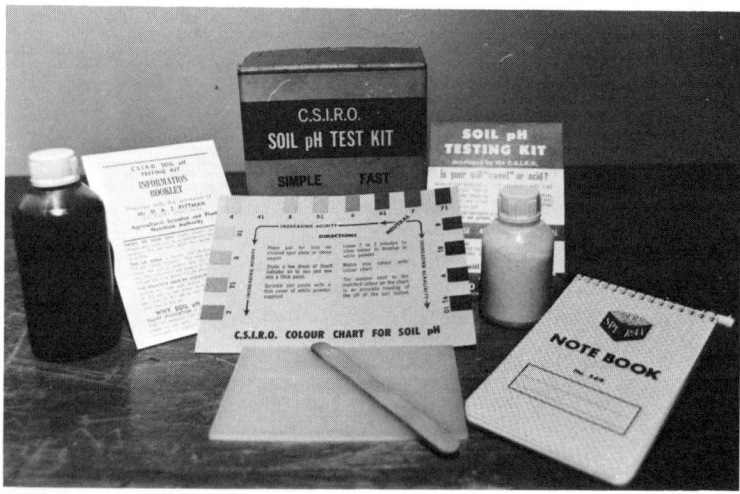

Soil Testing. Soil testing kit.

Soil Testing. Solution is added.

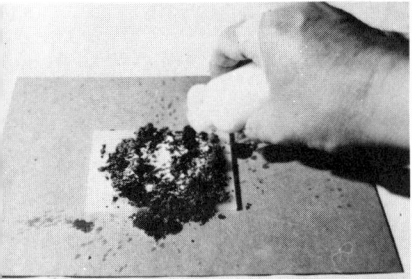
Soil Testing. Powder is added. Watch for change of colour.

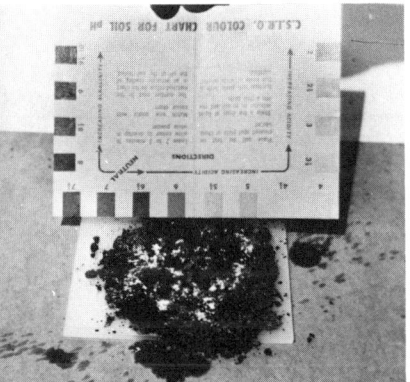
Soil Testing. Colour of powder is matched against colour on card.

Bulb Depth and Planting Considerations

It is logical to plant bulbs as near as possible to their accepted correct depth. However, to give a simple explanation of bulb planting depths is misleading and could lead to problems. Many factors determine the depth of the planting, such as whether the soil is light and sandy, or heavy. Generally, bulbs should be planted the right way up, at approximately their own depth in firm to heavy soil, and approximately twice their own depth in dry, sandy or light soil. This means that a bulb which is 50 mm (2 in) long should be

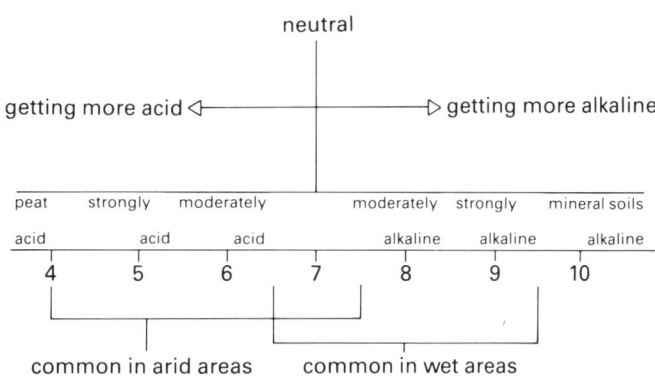

neutral

getting more acid ◁——————————————▷ getting more alkaline

peat	strongly	moderately		moderately	strongly	mineral soils
acid	acid	acid		alkaline	alkaline	alkaline
4	5	6	7	8	9	10

common in arid areas   common in wet areas

A pH Scale

planted 50 mm (2 in) deep in a heavy soil, and 100 mm (4 in) deep in a dry, sandy soil — the depth being measured to the top shoulder of the bulb. But even small bulbs generally need 50 mm (2 in) of soil above their crowns. It has also been suggested that bulbs be planted two or three times their largest diameter, the depth being to the top of the bulb, and not the bottom. Never lose sight of the fact that some bulbs like to be planted with their neck out of the soil. More planting details are given in the text.

Most bulbs are planted when they are in dormant stage. However, there are bulbs such as snowdrops (Galanthus sp.) and snowflakes (Leucojum sp.) that seem to recover and flower sooner if moved when in flower or immediately after.

The majority of bulbs prefer a well-drained soil, although you will see in the text that certain bulbs prefer the opposite.

Spacing Between Bulbs

A 'rule of thumb' method can be used, when in doubt, for spacing between bulbs when planting. It is based on bulb, corm, tuber and rhizome size, the height and spread of the plant, and the fact that bulbs will increase naturally. For example, commonsense tells that you will need a half wine-cask in which to grow a common agapanthus, and a 100 mm (2 in) pot in which to grow a snowdrop. Therefore, using these yardsticks, agapanthuses will need to be planted 450 mm (18 in) apart in the ground, while snowdrops can be planted 50 mm (2 in) apart. Generally, bulbs and corms are planted closer than tubers or rhizomes. Small bulbs such as snowdrops, grape hyacinths and lachenalias are planted 50–75 mm (2–3 in) apart. Medium-sized bulbs such as daffodils and tulips are planted 150–200 mm (6–8 in) apart, and large bulbs such as liliums 200–450 mm (8–18 in) apart. The same theory

applies to corms. Using popular bulbs, corms, tubers and rhizomes as a guide you will see how planting distances fall into a pattern. Another factor to be borne in mind is what kind of display you wish to create — informal, cut-flowers or show flowers. All these demand different spacings.

Although more detailed information is given in the text on planting depths and distances apart, I have found during my long gardening career that people are more confident if they can mentally establish limits and boundaries. Here, then, are usual spacings for some common bulbs.

Eranthis (winter aconite) (corm) 50 mm (2 in) apart

Ranunculus (tuber) 50 mm (2 in) apart

Galanthus (snowdrop) (bulb) 50–75 mm (2–3 in) apart

Lachenalia (bulb) 50–75 mm (2–3 in) apart

Muscari (grape hyacinth) (bulb) 50–75 mm (2–3 in) apart

Cyclamen species (variable tubers) 50–150 mm (2–6 in) apart

Babiana (corm) 75 mm (3 in) apart

Crocus (corm) 75 mm (3 in) apart

Crocosmia (corm) 75–100 mm (3–4 in) apart

Ixia (corm) 75–100 mm (3–4 in) apart

Bulbocodium (bulb) 100 mm (4 in) apart

Iris (small bulbs and Dutch) 100 mm (4 in) apart

Ixiolirion (bulb) 100 mm (4 in) apart

Leucojum (bulb) 100 m (4 in) apart

Narcissus species (bulb) 100 mm (4 in) apart

Nerine (bulb) 100 mm (4 in) apart

Acidanthera (corm) 100–150 mm (4–6 in) apart

Calochortus (bulb) 100–150 mm (4–6 in) apart

Chionodoxa (bulb) 100–150 mm (4–6 in) apart

Colchicum (bulb) 100–150 mm (4–6 in) apart

Gladiolus, small (corm) 100–150 mm (4–6 in) apart

Camassia (bulb) 100–200 mm (4–8 in) apart

Scilla (bulb) 100–200 mm (4–8 in) apart

Anemone (tuber) 150 mm (6 in) apart

Gladiolus, large (corm) 150 mm (6 in) apart

Hyacinth (bulb) 150 mm (6 in) apart

Iris (large bulb) 150 mm (6 in) apart

Ranunculus (tuber) 150 mm (6 in) apart

Tulipa, species (bulb) 150 mm (6 in) apart

Narcissus (garden) (bulb) 150–200 mm (6–8 in) apart

Tulipa (garden) (bulb) 150–200 mm (6–8 in) apart

Ornithogalum (bulb) 150–300 mm (6–12 in) apart

Begonia (tuber) 200–300 mm (8–12 in) apart

Lilium (small bulb) 200–300 mm (8–12 in) apart

Alstroemeria (Peruvian lily) (tuberous root) 225 mm (9 in) apart

Amaryllis (bulb) 225 mm (9 in) apart

Arum (tuber) 225 mm (9 in) apart

Iris (rhizome) 250 mm (10 in) apart

Lilium (large bulb) 450 mm (18 in) apart

Agapanthus (tuberous rootstock) 450 mm (18 in) apart

Dahlia (tuber) 600 mm–1 m (24–40 in) apart

Kniphofia (red-hot poker) (fleshy rhizome) 900 mm (36 in) apart

Strelitzia (bird of paradise) (rhizome) 900 mm (36 in) apart

Alpinia (rhizome rootstock) 1 m (40 in) apart

The spacing of tall climbing bulbs is governed by the spread of top foliage and flowers. If the spread is 1.2 m (48 in) then the bulbs should be spaced 1.2 m (48 in) apart.

Fertilisers

Newly purchased bulbs usually have enough nutrients stored within themselves to produce flowers during the first season after planting. However, bulbs, like most other plants, need the correct amounts of carbon, oxygen, hydrogen, nitrogen, phosphorous (phosphates), and potassium (potash), sulphur, calcium, iron, magnesium, manganese, zinc, copper, boron, chlorine, plus other essential elements and trace elements. The plant obtains carbon, oxygen and hydrogen either from the air (CO_2) or soil water (H_2O), and from these it manufactures carbohydrates.

The three major macro-elements are nitrogen, phosphorous (phosphates) and potassium (potash). Put simply, the plant uses nitrogen to manufacture green leaves, which in turn manufacture carbohydrates; phosphates to encourage root growth; and potash to encourage flower production. The leaves are the 'factory' for producing amino acids and giant, nuclear-protein molecules from carbon dioxide, water and soil-borne plant elements. A shortage of any of these will lead to a deterioration in the bulb's performance. Conversely, an excessive amount of any plant element will severely interfere with the bulb's metabolism and, indeed, may poison and kill it.

Bulbs, being large underground storage stems, are grown principally for the bulb, and not for the leaf (like an onion as opposed to a cabbage). They require a balanced (and that is an important word), low nitrogen, slow-acting, granular, complete, bulb fertiliser, containing the correct amounts of nitrogen, phosphates and potash. Being granular, it will leach through the soil to the roots gradually. Too much nitrogen will lead to lush leaves and few, and possibly, no flowers at all. However, complete lack of nitrogen will produce starved plants, with little leaf growth to manufacture the foods needed by the bulb.

Sulphur, calcium, iron, magnesium, zinc, manganese, copper, molybdenum, boron, chlorine and trace elements, are usually available in the soil, and you have no need to worry about these. Do not apply any without seeking expert scientific advice. An excess of one or some of the above could lead to a soil being poisoned for a long time, perhaps forever.

There are fertilisers compounded specifically for bulbs, and you would do well to check these out. Check also whether a particular fertiliser is suitable for the particular bulb, corm, tuber or rhizome you have in mind. Read the manufacturer's label carefully. Naturalised bulbs need a low nitrogen bulb fertiliser, but do not over-fertilise. The instructions on the packet of bulb fertiliser should tell how to use it.

Humus-producing materials such as well-rotted farmyard manure, well-rotted compost, peat and leaf mould do wonders in enriching the soil. The well-rotted farmyard manure must be well produced and collected. Cattle fed on nutritional foods will produce good manure, particularly if their urine is collected in the manure. Cattle fed on 'stomach fillers', such as straw, will not produce manure of the same quality. The same principle applies to horse manure. 'Starved' farmyard manure could lead to a yellowing of the plant's leaves.

Humus materials are of little use if the soil is waterlogged, therefore drainage, if in keeping with the bulb's nature, must be carried out. Fowl, pig and pigeon manure are too concentrated for bulbs, and these manures are best added to the compost heap during its construction. Fresh manure of any sort, with its 'free-ammonia' compounds, is poisonous to bulbs.

Composts–Soil Mixes (for seed sowing, cuttings and potting)

Growing bulbs in pots is popular but it is essential that the correct compost is used in bulb pot-culture. At the Rothamstead John Innes Institute in England during the 1930s, experiments were carried out to determine suitable, standard, reliable composts for professional growers. Prior to this every grower had his or her compost recipe. From these experiments were evolved the now famous John Innes Composts. These are soil-based composts. The formulas are as follows, with all parts being in bulk. It is assumed that the soil will be moderately heavy from fibrous, grass turf (not couch or kikuyu or similar stolon-rhizome grass), the turves stacked face downwards to rot; that the soil will then be riddled through a 9 mm (3/8 in) sieve; that the soil be sterilised; that the sand will be coarse, containing up to 3 mm (1/8 in) diameter particles; and that the peat will be granulated, dust-free moss or sedge.

Seed Mixture

2 parts medium loam; 1 part peat; 1 part coarse, clean, sharp sand. Add to each bushel (36.5 litres or 8 gallons) of the mixture, 42 g (1½ oz) of superphosphate, and 21 g (¾ oz) finely ground chalk or powdered limestone.

Cutting Mixture

1 part medium loam; 2 parts peat; 1 part coarse sand.

Potting Mixture

7 parts loam; 3 parts peat; 2 parts coarse sand. To each bushel (36.5 litres or 8 gallons) of this mix is added 21 g (¾ oz) of ground chalk or limestone, and 113 g (4 oz) of a mixture of 2 parts (by weight) hoof and horn meal, 2 parts superphosphate of lime, 1 part sulphate of potash. This is known as John Innes Potting Compost No 1.

For pots over 100 mm (4 in) diameter, double the quantity of chalk or limestone and fertiliser. This is known as John Innes Potting Compost No 2.

When growing vigorous plants in 200 mm (8 in) pots, treble the quantity of chalk or limestone and fertiliser. This is known as John Innes Potting Compost No 3.

Young seedlings will need to be grown in a compost which is intermediate between seed-sowing compost and the general potting compost.

When growing bulbs that hate lime, make sure to use a suitable acid, free-draining, bulb compost.

It is a good idea to check the pH of the

soil and compost, which should be neutral or slightly alkaline for bulbs that tolerate lime, or acid for those bulbs that do not like lime. Refer to the section on soil testing for information on soil pH.

The magic ingredient of the above mixes is the quality of the soil, and the wrong type of soil can lead to an inferior, unsuitable mix. John Innes Composts are used when they are relatively fresh and stable, that is up to four weeks old. The peat should be acid, granulated peat. Some peats can be alkaline when collected from bogs that are fed by limestone-hills water.

Soil-less Composts

The variability of the soil for composts led to experiments using composts that contained no soil, but used ingredients such as peat, sand, vermiculite, perlite, polystyrene chips and fertilisers. Cornell University in the United States was foremost in devising a host of soil-less composts for a wide range of plants. These composts have advantages over the John Innes Composts, and some disadvantages too. They are light, clean, easy to use and sterile. Some are compounded with long-lasting, slow-acting fertiliser. But they may dry out more quickly, and they do not have the buffer effect of the clay particles in the John Innes Composts. Nonetheless, they are extremely popular and widely used.

The main point about seeding, cutting and potting composts is that they must be suitable for the purpose you have in mind, in this case growing bulbs, corms, tubers or rhizomes. Check the instructions on the pack to see if the compost you are using is suitable for the bulb you are growing. Deal with a reputable supplier. Make sure the compost does not dry out, unless such a course of treatment is needed for the bulb's well-being, and check that the bulb receives the correct amount of fertiliser throughout its growing period.

Bulbs That Will not Flower

There are causes, other than pests and diseases, why bulbs do not flower:
1. Bulbs planted too deeply. Replant.
2. Immature bulbs. Have not reached accepted flowering stage (2-4 years). Have patience.
3. Excessive application of nitrogen fertiliser. Do not apply fertilisers containing nitrogen until the surplus has leached out of the soil.
4. Excessive general fertiliser application. Refrain from further fertiliser application until excess is leached out of the soil. Be more accurate with fertiliser application.
5. Compacted soil. Dig up and add sand, peat, well-rotted compost and so on to help break up the soil.
6. Poor growing conditions: too shady or too sunny; too many invasive roots from other plants; too wet a soil or too dry a soil. The remedies are obvious.
7. Climate too cold. The bulbs could possibly be grown in a greenhouse. You may be able to find a more sheltered spot in the garden. You may be able to create shelter.
8. Climate too hot. You may, if the bulb's nature permits, be able to create shade.
9. Bulbs have not received adequate previous summer baking. Bulbs could be in the wrong spot, or the climate may, generally, be too wet.
10. Bulbs are blind and produce only foliage. Separate clumbs and keep only those that will flower.
11. Unsuitable bulb storage: too cold or too hot; too damp or too dry. Remedy is obvious.
12. Overcrowding. Remedy obvious.

Bulb-Size Will Affect Flowering

The size of the bulb can make the difference between a magnificent show and a good one. Top-size bulbs are as mouthwatering to the gardener as a box of chocolates is to a hungry child. I shall quote measurements for a few well-known bulbs.

Crocus
Class 1 100 mm (4 in) circumference and above
Class 2 90–100 mm (3³/₅–4 in) circumference
Class 3 80–90 mm (3¹/₅–3³/₅ in) circumference
Class 4 70–80 mm (2⁴/₅–3¹/₅ in) circumference

Gladiolus
Extra large 140 mm (5³/₅ in) circumference and above
Class 1 120–140 mm (4⁴/₅–5³/₅ in) circumference

Class 2 100–120 mm (4–4⁴/₅ in) circumference
Class 3 80–100 mm (3¹/₅–4 in) circumference

Hyacinth
Extra large 190 mm (7³/₅ in) circumference and above
Class 1 180–190 mm (7²/₅–7³/₅ in) circumference
Class 2 170–180 mm (6⁴/₅–7²/₅ in) circumference
Class 3 160–170 mm (6²/₅–6⁴/₅ in) circumference
Class 4 150–160 mm (6–6²/₅ in) circumference
Miniatures 140–150 mm (5³/₅–6 in) circumference

Narcissus-Daffodils (Narcissus-dafodils are classified by their flowering capacity, not by circumference. The reason why certain daffodil displays are famous is that the gardeners use Extra Double-nose bulbs.) `
Extra Double-nose Class 1 D.N.1
Double-nose Class 2 D.N.2
Round-extra Class 1 D.N.3
Round Class 2

Tulips (early)
Top-size 110 mm (4⁴/₅ in) circumference and above
First-size 90–110 mm (3³/₅–4⁴/₅ in) circumference
Flowering-size 80–90 mm (3²/₅–3³/₅ in) circumference
Flowering-size 70–80 mm (2⁴/₅–3²/₅ in) circumference

Tulips (late)
Top-size 120 mm (4⁴/₅ in) circumference and above
First-size 110–120 mm (4²/₅ in–4⁴/₅ in) circumference
Flowering-size 100–110 mm (4–4²/₅ in) circumference

Hippeastrum
Top-size; Superior extra-selected 260 mm (10²/₅ in) circumference

Amaryllis
Extra strong 240–250 mm (9³/₅–10 in) circumference

All the above produce good flowers, but the larger ones give that extra show, which makes such a difference. There is a considerable difference in the price, as well.

3 Pests and Diseases

Bulbs, if grown in healthy soil and climatic conditions, will be remarkably free from disease and infection. If you know an area carries disease, do not plant bulbs until all the symptoms of the disease have been cleared up. If the area is fungus-prone — a muggy, skulking corner of the garden — you should not plant bulbs there either.

The same applies to soil pests. If an area is prone to soil pest ravages, avoid planting bulbs until you have got rid of the pests.

Prepare the soil for the bulbs well in advance and, generally, dig in what are known as humus materials — peat or well-rotted compost — but check text for the soil requirements of a particular bulb. Ensure that the soil drains well, as a waterlogged soil will rot most bulbs, although certain bulbs can be grown in waterlogged soils. Is it possible, if necessary, to raise the height of the bulb soil, say 150 mm (6 in) above the rest of the garden in that area, to help facilitate drainage?

Buy bulbs free from pests and diseases. This is commonsense, but many people are still looking for a bargain, which may turn out to be quite the opposite. A good horticultural product will command a good price. Check liliums particularly carefully, as the scales can be infected at the base, even though the bulbs still look perfectly healthy.

Care During Growing Period

Care must be taken during the growing period to ensure that the bulbs grow as near perfect as possible. Pay particular attention to their watering; never let them dry out during a dry spell, or waterlog them during a wet spell, unless the bulb requires this form of treatment. Check text for details. Attack diseases and insect pests when they are noticed.

Pests

Ants

Ants are the 'cowherds' of the garden, and are a problem because they round up and milk the aphids that attack the bulb foliage. The ants exchange protection for a fluid secreted by the aphids — a classic example of an insect protection-racket! There are various ant-killers on the market.

Incidentally, destroy insects only when they have become a major problem.

Aphids green (greenfly), black, root
These minute, green, black or various coloured, soft-bodied insects, cluster like scum on the tender shoots of the plant and, by lancing their proboscis (nose) into the tissue, draw out plant food. This does not necessarily damage the plant. The damage is done by the virus, which enters the wound and snakes through the plant's veins. Therefore, to control the aphids is to control the virus. There are many products available for spraying aphids. A golden rule is always to use the safest pesticide or insecticide. They are controlled by malathion or Pyrethrum or Dimethoate. Aphids are also kept in check by predators such as ladybirds, so check the spray and do not unnecessarily kill off the predators.

Root aphids attack the roots of many plant species, and the effect on the plant is similar to eelworm attack. Root aphids can be stored accidentally when the bulbs are lifted. Check bulbs. Root aphids are apparent as a floury, group-cluster among the bulb roots. Treat the bulbs with the manufacturer's recommended rate for bulbs of Dimethoate.

Bulb Mites

Bulbs can become infested with mites. You will need a magnifying glass to see these transparent insects as they cluster, like minute dewdrops, among the scales and roots. They attack, usually during storage time, and feed on the disc area where the roots emerge. If the mites damage the embryo roots the bulb will suffer. Mites attack tulips, amaryllises, daffodils, gladioluses, hyacinths and many more bulbs. Treat the mites once located, with Malathion or Dimethoate.

Earwigs

Earwigs lie in wait, and will venture out and eat the bulb foliage and flowers while you are asleep. They can disfigure a whole bed of flowers in a night. It is best to get rid of their hiding places to control them effectively.

Eelworms (nematodes): stem, bulb, leaf
These minute, sap-drinking insects are present in the soil, and their habit is to force entry into the sap stream and float about the plant, living on the liquid foods. Their presence causes distortion and death to the bulb and foliage. Eelworms will last for many seasons and, indeed, will stay alive in cyst form for many years without attacking the host bulb.

The first symptoms are distorted and twisted leaves, either at angles or curved down, sickle-fashion. This is followed by a yellowing, streaking of the leaves, which then turn sickly brown and die. Eventually the plant dies. The bulb, when cut into sections, displays a browning at the neck and base of the leaves.

Once you are sure it is eelworm attack, lift the bulbs and burn them. The bulbs can be lifted and treated by heat therapy, but this is not easy as you could inadvertently kill the bulbs. Seek professional advice. The soil in which the bulbs were growing will need to be fumigated to kill any eelworm left in the soil. Again seek local professional advice in identifying eelworm attack and its symptoms, heat therapy, soil fumigation and subsequent treatment. There are highly toxic nematicides, nematode-granules, for treating root-eelworm and highly toxic, or moderately toxic, systemic insecticide sprays for stem and leaf eelworm. However, once eelworm attack is suspected, call for professional opinion as to treatment. Do not take risks. Plant only healthy bulbs.

Mealy Bugs

As the name implies, these insects are coated with a mealy, floury skin, like specks of dirty, cottonwool. They attack flowers and can congregate along the veins of leaves, both above and below. Treat with Dimethoate systemic. Mealy bugs may be found on bulbs in storage. Treat with Dimethoate at the manufacturer's recommended solution for bulb dip.

Mice and Birds

Field mice will dig up liliums, gladioluses and other bulbs, take them away and eat them. Birds, such as sparrows, will eat the flowers of crocuses and other plants if they are hungry. These small animals are not a major pest in many urban home gardens.

Millipedes

These can reach plague proportions in some states. They are flattish creatures resembling undernourished centipedes, being smaller and greyish. They attack bulbs and seedling foliage. Treat with Carbaryl insecticide.

Narcissus Fly
Various species are grouped conveniently under 'narcissus fly'. Generally, the maggot that does the damage looks like fisherman's gentel maggot with a dirty white maggot body. It eats the tissue of the bulb, including the embryo flower. The adult fly lays its eggs amid the foliage when the foliage is dying back. The maggots emerge and eat out the heart of the bulb, reducing it to a soft, pulpy skin. Soft and sickly looking daffodil bulbs should be burnt.

Rabbits
Rabbits attack bulb foliage. Should they become a major pest, rabbit-proof the garden with a fence.

Red Spider
Red spiders are minute mites, not spiders, whose (webs) can only be observed under a strong magnifying glass. They attack the foliage drawing out sap. I have seen them attack and decimate plants outside in the garden, and in the greenhouse. The symptoms are greying of the foliage. Spray with Dimethoate systemic insecticide, or Kelthane.

Slugs and Snails
These crawlies will attack most young bulb foliage, particularly gladioluses, dahlias and liliums. They skulk in sheltered places under leaves or rocks or deep in organic, loose soil, then slide out when they think nobody is looking and attack the plant or its roots. Placing coarse, clean sand under and around the bulb when planting, will help deter the soft-bellied slugs. Slug and snail killer, obtainable from most nurserymen, will also help. You can help to eliminate slugs and snails by tidying up the garden and thus get rid of their hiding places.

Springtails
These are 'flecks of white dust' insects found on bulbs. They tend to spring like kangaroos when disturbed, particularly during bulb-lifting time. Springtails do not necessarily attack the bulbs, but feed on dead bulb tissue.

Stem Borer Grub
This pest usually attacks the taller, cane-stemmed plants such as cannas, hedychiums, alpinias, liliums and the like. Suddenly the leaves and sections of stem collapse. Inspection of the stem may find evidence of the borer. Spray with Dimethoate systemic.

Thrips
Thrips are summer visitors that crowd our plants, and their characteristic sign is the appearance of grey spots and streaks on bulb foliage, such as gladioluses, amaryllises and, to some extent cannas. Not only will they attack leaves, but they also attack the flowers before they open, thus disfiguring them. Remember to watch for the first signs of thrip damage — ghostly white-silver — on the leaves. Control by using Malathion or Pyrethrum sprays.

Weevil
Weevils will attack most green-leafed plants. They chew around the edges of the leaves leaving a scalloped, wavy, picotee-brown margin. Spray with Carbaryl.

Diseases
Bulbs are subject to many fungal diseases, and each bulb seems to have its own peculiarity.

Bud Rot (Cannas)
The leaves surrounding the buds are as black as chimney sweeps. The flower inside the bud is usually ruined. Select healthy, young stock to restore your canna beds and plant them well apart. They will soon clump together.

Bulb Rot
This disease is brought about by botrytis, or core soft rot, which goes on to infect the leaves with brownish spots and marks the edges with a reddish margin. Dig up the infected bulbs and burn them.

Botrytis
Botrytis attacks most bulbs and plants. Symptoms are the collapse of the leaves, which become carpeted with grey, felt mould. The grey turns into dry mud and cracks. Overcrowding can cause botrytis. Spraying with Zineb, plus a wetting agent, may help control it. Captan may also give some help. However, certain plants do not recover and inspection of the bulb reveals that the fungal 'sclerotia', like minute black knobs, is evident on the bulbs. These bulbs must be removed and burnt.

Fusarium Basal Rot
Fusarium rots the bulb disc area at the base of the bulb. The bulb turns muddy, brown-grey, and the fungal fingers feel their way through the scales and eventually strangle them. Wet soil and warm temperatures are ideal for this form of fungal attack. This is one good reason for ensuring that bulbs, if it is natural for them, have good drainage. Soil fumigants are available to combat fusarium basal rot, and to control rhizoctinia, pythium, didymella and verticillium. Get expert advice from the Department of Agriculture, as some soil sterilants are powerful and dangerous.

Gladiolus Scab
Gladioluses can be invaded by an army of diseases — botrytis, corm rot, leaf blight, fusarium and more. But the worst enemy is scab. This disease is caused by bacteria, and if action is not taken immediately to remedy the situation, a whole bed can be wiped out.

Start by inspecting the corms before planting. An infected corm will have circular, sunken, earthy spots. The spots will expand into watery-looking pools, and then to coal-scabs. The scabs are then attacked by virus or bacteria, and the corm rots.

Control is first to eliminate the cause — mostly it will be found to be bulb mite.

What happens to leaves of bulbs when they are kept in the dark, and then suddenly exposed to sunlight.

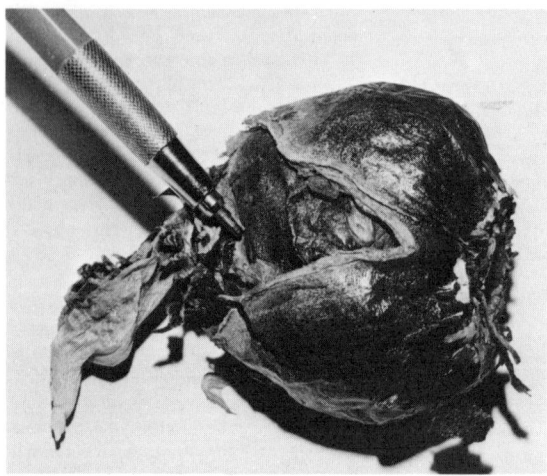

Neck-rot in hyacinth bulb.

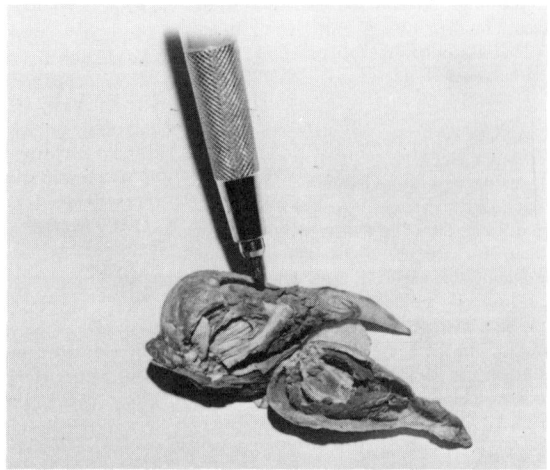

Bulb storage-rot in tulips.

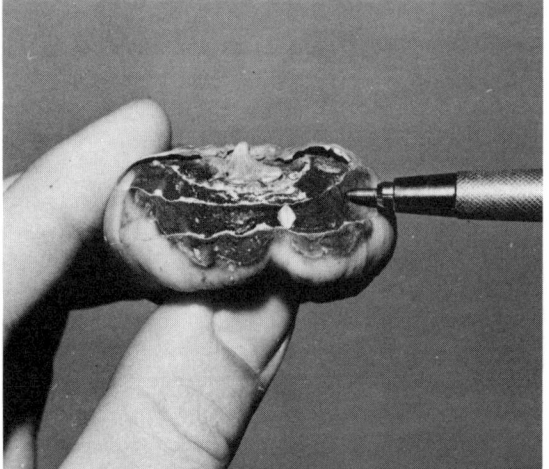

Gladiolus corm lesions revealed when outer skin was removed.

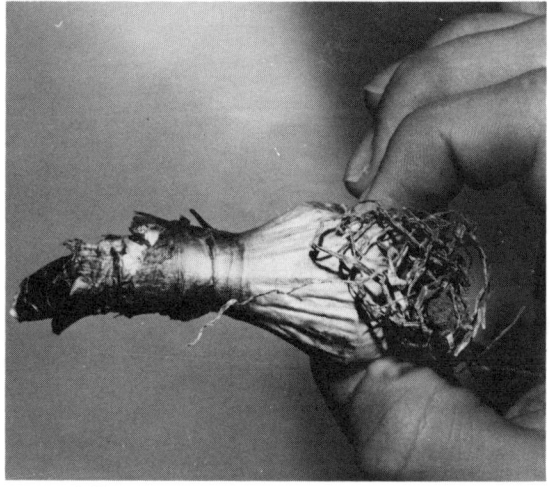

Narcissus bulb centre-rot.

Then treat with a suitable insecticide. Some greater control can be effected by spraying the diseased areas with Benlate or Bordeaux Mixture or similar.

Ink Spot
This disease attacks various bulb irises, and the symptoms are black blotches and smudges on the bulb tunic (skin). Eventually the disease enters the tissue and kills the bulb. Burn infected bulbs and treat those that are still healthy with a suitable fungicide.

Powdery Mildew
A ubiquitous disease which clouds all plants. The leaf fades to grey, and on closer inspection fungal spores are seen spinning fine cobwebs over the leaf. The disease is obvious on roses, and can be seen easily on bulbs. Control mildew by spraying the infected plant with Benlate.

Red Blotch
This causes stunting and distortion of the leaves. Spray with Zineb.

Rhizome Rot
Rhizome rot appears as earthy scar tissue over the rhizome and, when cut open, stands out as grey tissue in surrounding healthy, creamy tissue. Also it smells unpleasant, similar to rotted potatoes. Yellowing leaves may be an indication of the disease. You can cut out the infected parts and treat the bulb with Zineb but, more logically, you will have to improve the drainage.

Storage Rots
Many healthy bulbs are destroyed by storing them in slum-like conditions. Make sure the bulbs are clean and free from infection before storing, as one infected dahlia tuber, for example, will rot the rest of the tubers.

Tulip Break or Streak
A breathtaking feature of a tulip flower is 'breaking' — that is, a blatant red becomes more arrogant with stripes, bars and blotches. This colour break was considered of great value to the tulip growers in 'tulipomania' time. They did not know that 'breaking' or 'streaking' is a viral infection.

Many gardeners still prefer the variegation: green breathing over orange, red pencilling around the edge of a white flower, crazy paving gold patches, yellow flaming to orange, and so on. See more about 'bizarre' and 'Rembrandt' tulips later. But no matter how beautiful virus infected bulbs appear, they still have a disease — an incurable disease — and should be kept well away from healthy tulips.

Tulip-Bulb Temperature Poison
Stored tulip bulbs can be poisoned by temperature fluctuations. Hydrocarbon

gases build up within the bulb tissue, destroying the plant embryo. A cold snap will spark the tulip bulb into life. If followed by a warm spell it could lead to a build-up of these gases. This is where local expert advice on the best planting time for tulips is so helpful.

Tulip Fire
This disease rages through a bed of tulips like a bushfire. Botrytis disease is the fire-bug. The first symptoms are yellowish measles on the leaves, followed by star speckling which, in turn, is followed by the leaves hunching to the side of the leaf where the attack is severest. Flower colour fades and the leaves appear deformed.

Control this disease by digging up the bulbs carefully, so that no infected pieces fall to the ground to lie in wait for succeeding tulip crops. All infected bulbs should be destroyed, and the area not planted with tulips for three to five years — some experts say three years, others five years.

Chlorosis
Bulbs, if short of a certain chemical, will exhibit strange characteristics. One most dramatic manifestation of nutrient shortage is chlorosis, or 'white-yellowing' of the leaves. A deficiency of iron, magnesium or manganese will result in white-yellow plants. it could be that there is too much lime in the soil and the other elements are being driven out. First obtain local expert opinion as to the cause, and bear in mind that you can obtain chelated iron, magnesium and other lime-resistant elements

Virus Diseases
It is reasonably true to state that all bulbs will be attacked by virus diseases sooner or later. Virus can cause streaked, mottled, stunted or distorted dead areas, broken flowers, and death to the plant. What one has to do is to control the vectors that spread the disease — aphids and the like, which puncture the leaves to get at the sap, and leave a wound to be invaded by the virus.

Slugs, snails, earwigs and other crawlies leave wounds. Rough handling by you, your children and animals could also cause wounds, which would be attacked by virus. Indiscriminate forking around the bulb clump is damaging. You may wish to remove a weed that is stubbornly entangled around the bulb clump. Remember the bulb when removing the weed!

In Summary
Bulbs can be attacked by more pests and diseases than those mentioned above. There are offices in many states where the home gardener can seek advice on gardening problems, therefore if you are not sure about a point seek professional guidance. Don't take chances with fungicides, pesticides, insecticides, herbicides (weedkillers), nematicides or soil sterilants. There are state laws governing the use of certain chemicals. Check the pack.

1 Read the label on the container to ensure that the fungicide, pesticide, insecticide or herbicide you are using is suitable for the job. Certain plants are allergic to certain insecticides, pesticides and fungicides. Certain insecticide or fungicide sprays may be flammable. *Read the label.*

2 Read the label and follow the directions precisely.

3 Use the least toxic substance that will do the job.

4 Some herbicides are volatile, as are some fungicides, pesticides and insecticides. Off-target damage can be considerable — as well as the cost to you. Use the least volatile material to reduce spray drift. If feasible, use coarse nozzles to help reduce spray drift.

5 Choose a cool, calm, windless day to apply the spray.

6 Keep children and pets well out of the way.

7 Certain insecticides and fungicides will kill fish, or may poison bees.

8 Certain insecticides and fungicides should not be used anywhere near edible crops such as vegetables and fruit, especially at harvest time. *Read the label.*

9 Lock up the chemicals after use, in a sensible spot that is not too high, so that chemicals could fall into your eyes, face and mouth, or too low for children to find easily.

10 Wash your hands and face after use.

11 Do not use fungicides, pesticides, insecticides or herbicides unless you have to. Remember that the pest's natural predators will substantially control it.

Herbicides, pesticides, insecticides, nematicides and fungicides are available in your local home gardeners' plant nursery. Some are well known by their brand name while others are known by their 'active constituent' (ingredient) name. Some of the most common are listed below. Check the pack for details on what each contains, safety precautions and its action on pests or diseases:

Benlate (benomyl); Bordeaux Mixture (copper sulphate-hydrated lime); Captan (captan); Karathane (dinocap); Mancozeb (mancozeb); Zineb (zineb); Thiram (T.M.T.D). Carbaryl (carbaryl); Derris (rotenone dust); Endosulfan, Thiodan (endosulfan); Kelthane (dicofol); Lebaycid (fenthion); Malathion (maldison); Metasystox (demeton-S-methyl); Nemacur (fenamiphos-phenamiphos); Rogor (dimethoate).

4 What Bulb Is That?

There follows an alphabetical list of species, varieties and cultivars of well-known, and not so well-known, bulbs, corms, tubers and rhizomes, many of which are grown extensively in Australia and New Zealand.

A large proportion of the world's bulbs will grow in at least some parts of Australia and New Zealand, these two countries providing environments ranging from sub-antarctic to tropical. Bulbs, corms, tubers and rhizomes, being among the most popular and colourful plants in the world, are deservedly popular and will repay the work put into growing them.

It may be that some of the bulbs, corms, tubers and rhizomes described are not available at the moment, or may never be available due to circumstances beyond the control of Australian and New Zealand gardeners. However, I consider it important that you should know about them.

Botanical Names

Botanical names are a necessary evil. Common names can change from town to town, from state to state and from country to country. Botanical names, on the other hand, are accepted world-wide. The original name given to a plant, when found, will be adopted — this is why they seem to change so regularly. Suppose we know a lilium botanically as *Lilium brownii* and then we find an earlier reference calling it *Lilium jonesii*. We then adopt *Lilium jonesii* and all the botanical references are changed accordingly. Then somebody discovers an older reference calling it *Lilium smithii*. We then have to adopt the name *Lilium smithii* and change the reference again. The oldest known botanical name will be the accepted one.

Pluralising Latin

Most know the plural of 'gladiolus' as 'gladioli': a bunch of gladioli. Right? Wrong! There is only one genus of gladiolus and more correctly it is 'a bunch of gladiolus'. You can have a bunch of gladiolus species. However, modern jargon has adopted 'i' or 'es' as a plural to names such as gladiolus. Which shall we adopt? Gladioli sounds good, as does narcissi, but what about agapanthi and loti? Do two platypus make one platypi?

I thought of dispensing with plural

endings, but 'a clump of nerine' does not convey the same feeling as 'a clump of nerines'. I have opted for the more modern 'es' ending — gladioluses, narcissuses, agapanthuses, lotuses — throughout the book. The plural of words such as cyclamen I have left as cyclamen (or should it be cyclaperson?). Where I have considered a plural 's' is necessary to make 'modern' sense I have added one.

Pest Plants

Having been a State Authorised Pest Plants Officer, I know that certain bulbs have become pest plants. This is why I advocate that any surplus bulbs should be destroyed, and not thrown away. Should you wish to seek information on bulbs you consider as pests, bear in mind that many councils have weeds and pest plant officers on their staff. Each state has places of reference for pest plants. Check the state offices reference in the phone book. *Australian Capital Territory*: Department of Territories, Local Government and Agriculture Section. *New South Wales*: Noxious Plants Advisory Committee of NSW, Department of Agriculture, Sydney. *Northern Territory*: Department of Primary Production, Darwin. *South Australia*: Pest Plants Commission, Adelaide. *Tasmania*: Department of Agriculture, Hobart. *Victoria*: Metropolitan Melbourne, Department of Agriculture, Garden Advisory Service; Outside Melbourne: Vermin and Noxious Weeds Branch, Department of Forests and Lands, Melbourne. *Queensland*: Weeds Section, Stock Routes and Rural Lands Protection Board, Brisbane.

Achimenes You will find these colourful, greenhouse, horn flowers cascading from hanging baskets. The small rhizomes are placed inside the compost of the hanging basket and, eventually, will grow through the wire and waterfall down the outside. Their flowers can be had in pink, blue, purple and orchid-pink. *Achimenes longiflora* **(plate 13)**, *A. heterophylla*, *A. pedunculata* and their hybrids are most frequently grown.

The small, scaly, grub-like rhizomes can be planted 13–25 mm (1 in) deep from July to October in boxes or trays of African violet compost, which is readily available from the garden centre. Keep achimenes in as warm an atmosphere as you would keep African violets. As they are frost sensitive they will not tolerate a temperature below 15.5°C (60°F). Water sparingly until the rhizomes have produced

shoots, as any excess watering at this stage could cause rhizome rot.

Move the plant nearer the light once the rhizomes have developed leaf shoots, but not into direct sunlight as this causes scorching. When the leaves have reached 75 mm (3 in), transplant them 3–50 mm (1½–2 in) apart into their permanent homes or, singly, into 75 mm (3 in) pots, or a few 38–50 mm (1½–2 in) apart, in a 125 mm (5 in) or 175 mm (7 in) pot. There is no need to shift those planted in hanging baskets.

Feed the plants as they grow with an achimenes liquid fertiliser. Achimenes will grow well in a shaded place during the summer, and will produce their characteristic cascades of coaching horns.

The rhizomes can be collected, dried naturally, and stored for the next planting season, when the achimenes have finished flowering. Most of the root can be used to grow flowers if the section has a growing bud.

(Achimenes: 'cheimaino'-tender; or after Achaemensis, a Persian king, and ancestor of Cyrus the Great. *Family*: Gesneriaceae)

Acianthus (see *Orchids*).

Acidanthera *Acidanthera bicolor* grows in swarms in Ethiopia and areas to the south, and resembles three-winged butterflies, the centres being blotched, spotted or even starred. They are available in white and orange, and shades of these. They grow to 600–900 mm (24–36 in) and bear growth characteristics similar to the gladiolus.

The recent white cultivars (garden-bred varieties) of *A. b. murielae*, which can grow as tall as 900 mm (36 in), have a sweet fragrance that makes them desirable in a garden.

Acidantheras can be used as cut flowers, as their lifespan in water is as good, if not better, than gladioluses. Their perfume is a household joy on dull days.

Plant the corms 75–100 mm (3–4 in) deep, and 100–150 mm (4–6 in) apart in a well-drained, friable soil during spring. Keep them watered during hot weather, but never allow them to become waterlogged. Treat them as you would gladioluses, but *do not* remove the spawn corms from the roots.

(Acidanthera: from the Greek 'akis', a point, and 'anthera' meaning slender stamens. *Family*: Iridaceae)

Agapanthus (African lily of the Nile) This South African, summer-flowering, bulbous plant dominates many Australian gardens because of its explosion of sky blue flowers and the ease of growing it. Agapanthuses are available in varying shades of blue; a white form is also available. The blues range from thunder blue to pale porcelain. A variegated leaf form, *A orientalis* **(plate 1)**, 1.5–1.8 m (5–6 ft) tall, with its balloon head of flowers on one spike — indeed a single head can contain scores of flowers — is most popular. The seedheads left after flowering are small, green explosions, and are used by florists, as the unusual shape of the seed cluster can lend itself to unusual decorations. Many prefer the smaller 600 mm (24 in) *A. africanus*.

Agapanthuses can be grown in most garden soils, and can be seen in all states. They like a rich soil, adequate water when they are growing well, and partial shade.

Prepare the beds by deep digging and incorporating well-rotted manure or acid peat. An artificial fertiliser can be added at the manufacturer's recommended rates. Planting can be done autumn–winter.

Divide the clump into smaller pieces and place them 450 mm (18 in) apart. They will soon grow together, but let them crowd for a few years before dividing again. Being uprooted and forcibly moved to another home causes shock, and the agapanthus will reflect this by not flowering the first year. Give the plant time to settle down.

Agapanthuses will bloom in early summer, and as backdrop for *Amaryllis belladonna* they combine to become a perfection of pink and blue. We have mentioned *A. orientalis*, but also available are its varieties: *A. o. albus* (plate 14), *A. o. pallidus*, and *A. o. flore pleno*. The deciduous *A. campanulatus*, the bell-flowered agapanthus, is 900 mm (36 in) tall and produces 450 mm (18 in), narrow foliage. The flowers have that turned-back, tipped, bell-shaped characteristic.

Clumps of agapanthuses are a haven for all manner of pests, including mice and rabbits, but mostly they will house snails and slugs. You can clear snails and slugs using slug bait.

(Agapanthus: 'agapas', love; 'anthos', flower. *Family*: Liliaceae)

Albuca Albucas originate from south and tropical Africa. The white, yellow or greenish white flowers, usually with a green or red or brown stripe down each outer petal, can be erect, or pendulous or nodding. They are 25–38 mm (1–1½ in) long and about 12.5 mm (½ in) wide. They appear in November on naked, loose-flowering, unbranched spikes, or racemes, like ornithogalums, from broad, grass-like foliage.

The flowers of *A. nelsoni*, borne on a 1.5 m (60 in) tall spike, are upright, clear white, 38 mm (1½ in) long, carried on 50–75 mm (2–3 in) stalks emanating from the spike. The three inner petals are touching and seemingly joined as a tube, while the three outer, brown-keeled, pointed petals are separate and spread away from the inner petals like a split, crinoline skirt. The foliage is dark, blue-green, 50 mm (2 in) wide, and pointed.

A. canadensis (*A. minor*) has pendulous, 25 mm (1 in) long, yellow, outer petals, and shorter, pale, inner petals, seemingly fused as a tube, like a fuschia flower. The blooms are borne on 400–900 mm (16–36 in) stout flower spikes. The broad-based, grey leaves, 150 mm (6 in) long, 25 mm (1 in) wide, taper abruptly to a point.

Plant the flattish-round 50 mm (2 in) diameter bulbs in autumn. There are about eighty species to choose from. *A. humilis* is also grown.

(Albuca: 'albucus'; album; original sp. has white flowers. *Family*: Liliaceae)

Allium (ornamental onion) Ornamental onions can be extremely attractive flowering plants. However, some of the more vigorous species can get out of hand rapidly. Many of the flowering garlics have become great pests and have caused problems to farmers.

Bulbs are available for planting during the autumn. They are greedy and like a rich soil. The depth of planting depends on bulb size, but the larger ones can be planted 100 mm (4 in) deep, and the smaller ones 50–76 mm (2–2½ in) deep, 100–150 mm (4–6 in) apart.

The bulbs will flower during early summer, and carry on blooming for some time. The heads are powder-puffs of white, yellow, red, blueish or purple.

A. christophii (*A. albopilosum*), a ball of vivid, violet flowers. *A. caeruleum* (*A. azureum*), the blue garlic, a constellation of star flowers. *A. flavum*, rich yellow flowers. *A. moly*, the golden garlic, buttery yellow blooms above broad leaves — grows up to 625 mm (24 in). *A. neapolitanum*, the daffodil garlic, white flowers, tall stem, no smell. *A. sphaerocephalon*, dark purplish, tall — 900 mm (36 in). There are more.

(Allium: Latin for garlic. *Family*: Liliaceae)

Alocasia (spoon lily) *Alocasia macrorrhiza* has largish, heart-shaped foliage, 1–2 m (40–80 in) tall, bearing spoon-shaped, 'calla-lily' green flowers, which are sweetly fragrant. They are frost sensitive and should be protected. Plant the fleshy roots in spring, in a rich soil in a shady area, and keep them well watered during dry spells. Plant them 900 mm–1.2 m (36–48 in) apart. Can also be grown in large containers.

(Alocasia: variant of colocasia. *Family*: Araceae)

Alophia (see *Herbertia*)

Alpina (wild ginger or shell flower) *Alpinia speciosa* will grow to 1.5 m (60 in), and twice that in warm climates, and can become a problem. Shell flower or wild ginger once it becomes established, will spread rapidly and one is forced to ask whether it has a place in the small, suburban garden. Wild ginger should be placed in a sheltered site where it can be controlled. The fleshy roots will soon spread, and have to be contained. The bell- or shell-shaped, spicy, fragrant, spiky, white-orange and yellow waxy blooms are a joy. Other species are: *A. formosana* 1.5 m (60 in) tall; *A. zerumbet*, 2.4–3 m (8–10 ft); *A. coerulea* 600–900 mm (24–36 in) tall.

Alpinia rhizomes can be planted in late spring 50–75 mm (2–3 in) deep, preferably in a rich organic soil, say 1.2 m (48 in) apart. They need copious moisture when growing well, and regular thinning to keep them in bounds.

(Alpinia: in honour of Prosper Alpino, Italian botanist. *Family*: Zingiberaceae)

Alstroemeria (Peruvian lily) One can visualise the colourful Incas in these beautiful summer-flowering bulbs, with their host of yellow, cream, fawn, orange and red shades, produced on stems 600–1200 mm (36–48 in) tall. Alstroemerias can be had in streaked, self or spotted markings. ('Self' means 'all one colour'). They are easy to grow.

Dig the soil well, incorporating peat or compost, and place the roots 100–180 mm (4–7¼ in) deep and 225 mm (9 in) apart. Leave them to flower undisturbed for years, when they can be dug up, separated and replanted. They like to receive full sun. Keep the tuberous roots in check, making sure they do not escape and become a weed nuisance. Burn or compost any surplus bulbs.

Alstroemeria aurantiaca, the common Peruvian lily, and its cultivars are grown most There is a strain of Peruvian lilies known as Ligtu hybrids, which have a magnificent, high blood-pressure colour range. The Peruvian parrot lily (*A. psittacina*) can be recognised by its deep red, delicately green-tipped foliage.

(Alstroemeria; in honour of Baron Clas Alstroemer, friend of Linnaeus, the great botanist. *Family*: Amaryllidaceae)

Amana *Amana edulis* (syn. *Tulipa edulis*) is a Japanese–Chinese early flowering bulb, with spring onion-like foliage and small, white, star-shaped, 'tulip' flowers. Needs a well-drained soil. A collector's plant.

(Amana: Japanese name. *Family*: Liliaceae)

x Amarcrinum This bulb is a cross between an *Amaryllis belladonna* and *Crinum moorei*, and is known as an intergeneric hybrid. The resultant plant is beautiful, carrying many of each parent's characteristics, particularly *A. memoria-corsii*. A collectors' plant.

(*Family*: Amarylldaceae)

x Amarine An intergeneric cross between an *Amaryllis belladonna* and *Nerine bowdenii*. Attractive magenta-pink, or flush pink or rose-carmine forms are available. *A. x tubergenii* a collector's plant.

(*Family*: Amaryllydaceae)

Amaryllis (belladonna lily) The belladonna in *Amaryllis belladonna* (plate 2), meaning beautiful woman, and that sums up this most beautiful of plants. The 450 mm (18 in) tall, belladonna lilies can be grown practically anywhere. Each year orchestras of lily-like trumpets, 125 mm (5 in) long, by 75 mm (3 in) wide at the mouth, pink and white-throated, in clusters of up to ten or more, appear before the foliage.

Belladonna lilies can be grown naturalised in beds of their own, or they can be planted along garden borders. I have seen these lilies massed in light woodland and in full sun, but overall a sunny spot, not too exposed, would be preferred.

Amaryllis bulbs are huge and should be planted, after the flowering period, just below soil surface (25–50 mm) in the warmer parts of Australia. In the cooler parts plant them with their necks just protruding. Space them 225 mm (9 in) apart.

Amaryllis prefer a deep, well-drained soil, although they will grow in most garden soils. They like to remain undisturbed, and that is why they seem so effective in untidy gardens.

A mulch of compost over the bulb area each year will help to feed the bulbs. Cut down the stalks once the flowers are finished and tidy the beds. Let the foliage die off naturally.

There are many sweetly scented forms available, each with an exciting colour variation of pink, white, white-cream, red, red and white, and pink and white. *A. b.* 'Hathor' is white; *A. b.* 'Cape Town', deep rose pink; *A. b.* 'Rubra', deep pink-red; *A. b.* 'Windhoek', pink, with a white throat.

(Amaryllis: Greek feminine name. *Family*: Amaryllidaceae)

Ammocharis *Ammocharis coranica*, the ground lily, has a 150–200 mm (6–8 in) diameter, agapanthus-like cluster of up to twenty, rosy red, fragrant, star-shaped flowers, borne on 150–250 mm (6–10 in) dumpy stalks. The flower stems rise out from elongated, flat, withered-tipped, two-tiered (aerial and ground-hugging), blue-green, rosettes of leaves. These leaves are 450 mm (18 in) or more long and 37.5 mm (1½ in) wide. The large bulb, up to 200 mm (8 in) across, should be planted in spring, just below ground level, 300 mm (12 in) apart. It requires full sun and a well-drained soil. Leave undisturbed for a few years. Those who experience wet summers or frost-prone winters would do better by planting the bulbs in large 200–225 mm (8–9 in) deep pots and grow them as pot plants. The pots can be sunk into the sand-base of a bulb frame during summer to keep the roots moist, but lifted when the top tier of leaves has died back to the bulb, even though the basal leaves stay on the plant and extend each year for a number of years. Store the pots in a dry spot during the dormant season as they rot if left outside.

(Ammocharis: 'ammos', sand; 'charis', beauty. *Family*: Asteraceae)

Anemone (windflower) Referred to in the Bible as the 'lilies of the field'. I spent many months in the Middle East and can vouch for their wild magnificence, along with the poppies, daisies and viper's bugloss. Anemones are magnificent as they flower on 150–375 mm (6–15 in) stems so profusely in Australia and New Zealand.

Anemone coronaria is most planted. it has 75–125 mm (3–5 in), poppy-like blooms in red, lavender, purple and blue. The 'De Caen' single, and 'St Brigid' (plate 19) semi-double strains, are famous for their bright displays of red, cherry, white, violet and blue. *Anemone blanda*, with its deep blue flowers, can be found growing wild in Mediterranean countries. *Anemone apennina*, a native of Italy, fills the hills with blue in spring.

Plant the 12.5 mm (½ in), irregularly shaped anemone tubers between March and May, the right way up, 150 mm (6 in) apart and 37.5–50 mm (1½–2 in) below ground level. Do not plant the tubers upside down, as it is deleterious to subsequent growth. Check to find the root basal area, which is usually bristly. If in doubt plant the tubers on their side. Do not plant them in a soil that has had fresh manure incorporated. Anemones will flower from early spring onwards. They grow well in warm, moist conditions and are much suited to Queensland. Once the foliage dies, lift the tubers and store them.

Sow anemone seed in late summer for spring flowers. First dig in well-rotted compost and then break down the soil to form a fine tilth. The soil and subsoil should be reasonably consolidated before sowing the seed, otherwise it could settle and smother the seed. Spread a complete balanced fertiliser and rake it lightly into the surface. Fluff out the seed and scatter it over the seedbed, or place it in rows. Lightly cover it with sand or sieved, sandy soil. Water the seed gently, but well, with a fine rose on the hosepipe. From then on water the seed when it needs it. Prick out the seedlings, when they are large enough to handle, into beds 200 mm (8 in) apart, and 250 mm (10 in) between rows.

(Anemone: 'anemos', wind.

Family: Ranunculaceae)

Anigozanthos (kangaroo paw) Kangaroo paws are native Australian bulbs. A. *manglesii* has intense green and red, velvety flowers and is the flower emblem of Western Australia. A. *bicolor* also has fine, furry red and green blooms. A. *humilis*, the cat's paw, has velvet orange and yellow flowers. A. *flavidus* has green and yellow flowers. A. *viridus* also has intense, velvet green flowers. There are many lovely hybrids produced from these species. *Macropidia fuliginosa*, the so-called black kangaroo paw, has a black flower base and velvet yellow tube. Flowering time can vary from spring to autumn depending on species.

Their height can vary from 900 mm to 1–8 m (36–72 in), although out of their natural environment they may not reach full height. Many find difficulty in growing kangaroo paws, and they do tend to look a bit sick if in the wrong soil and location. In their natural state the thickish, red rhizomes do not last for many seasons, but kangaroo paws readily regenerate from seed. Plant the roots in autumn, after flowering, 300 mm (12 in) apart, in a warm, sheltered, well-drained, sandy loam. Some, such as A. *flavidus*, are hardier than others.

(Anigozanthos: expand; flower open to base. *Family*: Amaryllidaceae-Haemodoraceae)

Anoiganthus (A. *breviflorus* now included in *Cyrtanthus* species.)

Anomatheca (See *Lapeyrousia*.)

Anthericum (St Bernard lily) *Anthericum liliago*, the St Bernard lily, like the St Bernard dog, is named after the patron saint of travellers in the Swiss alps. Long 600 mm (24 in) scapes (stems) of 25 mm (1 in), lily-flowered, star-shaped, white blooms arise out of 300–450 mm (12–18 in) tall, grass-like foliage. A. *l. major* is a larger variety. They are usually grown as pot plants in the shadehouse or similar, but can be grown in herbaceous borders or rock garden areas, provided the site is not too exposed to hot sun. Propagate by division of tuberous roots, planted 150 mm (3 in) deep in heavy soil, 300 mm (6 in) deep in sandy, loose soil, and 450 mm (9 in) apart.

(Anthericum: Greek name for wheat stalk. *Family*: Liliaceae)

Antholyza *Antholyza ringens* produces 500 mm (20 in) tall stems, that are flowerless at the top on which honeyeaters perch to pollinate the flowers. The lateral, one-sided flower spikes, bearing eight to twelve, red, green-tubed, freesia-like blooms are 100–150 mm (4–6 in) long, and flower just above ground level. The leaves are corrugated, sword-shaped, 100–200 mm (4–8 in) long and 7 mm (¼ in) wide.

This is not a well-known corm, although it is attractive. Plant 150 mm (6 in) deep in sandy loam, and 100 mm (4 in) deep in heavy loam. A good pot-plant, but use a large 225 mm (9 in) pot, as the corms burrow deep into the soil.

(Antholyza: flower-rage; flower resembles enraged beast. *Family*: Iridaceae)

Anthurium *Anthurium* species and cultivars are from the same family as the calla lily. Their shiny, 'plastic' flowers (spathes), and straight or curly, cream, lemon, pink or red spadix (centre seed spikes) look unreal. They are tropical, and in most areas are greenhouse plants. They are grown in pots, and have to be treated accordingly. There are hundreds of species available, mostly A. *andraeanum* (**plate 20**), A. *cordifolium*, A. *crystallinum*, A. *hookeri*, and A. *scherzeranum* are grown. Of these, A. *andraeanum* and A. *scherzeranum*, and their cultivars, are most popular as cut flowers, lasting for a considerable time in water. The colour of the flower (spathe) varies from Grenadier Guard scarlet to pillar-box red to maroon to rose to pink to variegated reds, pinks, greens and white. Many species are grown for their arrowhead foliage.

The winter temperature of anthuriums must not go below .15.5°C (60°F). You should buy seedlings or established plants to begin. Plant the bulb with its base at ground level, with the ground roots seated firmly in the compost, taking care not to damage the roots. The plant will send out aerial roots, which must not be covered with compost, although some growers wrap sphagnum moss around them to get the roots to travel down into the pot compost. Anthuriums require a warm, buoyant atmosphere, and shade from the hot sun is essential — in effect conditions similar to those they would encounter in their native tropical America. They require watering during winter, but too much water during the bulb's dormant stage will rot it. They need high summer temperatures and high humidity.

Anthuriums like a peat, humic compost to which grit-sand has been added to improve drainage. They can be fed with a weak, organic, liquid fertiliser when they are growing well. Division or separating the roots can take place in late winter, just before the plant starts into active growth. Great care must be taken not to damage the roots when dividing them. Seed, which can be difficult to grow, is sown in a peat-moss compost, in a propagating frame with a 23.8–30°C (75–85°F) bottom heat.

(Anthurium: flowering tail-spadix. *Family*: Araceae)

Argyropsis (see *Zephyranthes candida*)

Arisaema *Arisaema candidissimum*, A herbaceous perennial from western China, is an early summer-flowering member of the Arum family, and produces mostly green, striped, hooded-funnel, flower spathes. It has a particularly lovely 75–100 mm (3–4 in) spathe, with white and pink flush, open or semi-hooded, with an acutely pointed apex that is a 25–50 mm (1–2 in) twisted, wiry tail, borne on a 150 mm (6 in) stem. The leaves reach from 300 mm (12 in) to 750 mm (30 in) in height, but do not achieve maturity until after the spathe has matured. The leaf stalks have three leaflets, which are widely oval, 75–200 mm (3–8 in) long, and the same wide. The centre, pencil-like, true-flower spadix is white to creamy yellow. Arisaemas thrive in damp places. Plant the 37 mm (1½ in) thick tubers in pots, or just below ground level, 225 mm (9 in) apart.

(Arisaema: 'arum' — blood-red.) *Family*: Araceae)

Arisarum *Arisarum vulgare*, another spring-flowering Arum family plant, is a small and unusual, funnel-shaped, hooded, striped, purplish, tubular-flowered plant, similar to our native *pterostylis* hooded orchid. It prefers a woodland soil, or sandy soil with added leaf-mould. Propagate by dividing the tubers, usually in spring. Grow divisions in pots to establish them. A. *proboscoideum* is also well known, as is A. *simmorrhinum*. Collectors' plants.

(Arisarum: ancient name. *Family*: Araceae)

Aristea Of the three well-known evergreen aristeas — *A. eckloni, A. thyrsiflora,* and *A. macrocarpa* — *A. eckloni* is most popular. The lance-like leaves are 200 mm (8 in) tall. A mass of 500–600 mm (20–24 in) stems arise from the foliage bearing 10 mm (½ in), open, ultramarine blue flowers. The flowers open during the day and close at night. *A. thyrsiflora* is a giant aristea, up to 2 m (6½ ft) tall. Divide aristea roots in autumn and space them 300 mm (12 in) apart in a moist, rich soil, in a sheltered but open site, although they will tolerate partial shade. They are also used as marginal, waterside plants.

(Aristea: point-rigid tips of leaves. *Family:* Iridaceae)

Arthropodium (rock lily; New Zealand rock lily) *Arthropodium cirrhatum* (**plate 21**) is so easy to grow in temperate Australia that one wonders why more people are not interested in this summer-flowering member of the lily family. Perhaps it is because the plant has to become well established before it flowers. It has thick, fibrous roots. The foliage is bright evergreen and strap-like, similar to grass or lilium foliage. The 25 mm (1 in) long, white flowers, borne on 600 mm (24 in) stems, are like large snowflakes, although the centre yellow and pink stamens give them a more summery feel. The plant is hardy and will grow in sun or some shade, in a well-drained soil. Propagation can be by division of the roots, or seed. A useful pot-plant. *A. candidum* is also grown.

(Arthropodium: Greek, jointed foot; jointed flower stalk. *Family:* Liliaceae)

Arum There are ten popular species of arum. Many of the so-called arums are not really arums, but species of the following genera: *Arisaema, Calla, Dranunculus, Eminium, Richardia* and *Zantedeschia.* Arums are found in North America, Europe, North African and China.

Arums look like ornamental funnels, and the black-red-throated forms are particularly attractive, if somewhat retiring amid green undergrowth. The funnel flowers are not true flowers, but are known botanically as spathes. The true flowers are contained on the spike-shoot (spadix) found in the centre of the spathe.

Arum italicum has large, arrow-shaped, marbled leaves, with lime green spathes. *A. maculatum,* Lord and Ladies, has a white and purple spathe. The popular *A. palaestinum* (**plate 22**), the black arum, has dark reddish spathes, with black seed spike (spadix) and large arrow leaves. Its scent can be overpoweringly offensive. *A. discordis,* mottled black-yellow with mottled spathes, smells like rotten meat.

The tubers should be planted in a rich, moist soil between March and May for flowering during the spring. The depth to plant varies with the size of the tuber as some species produce larger tubers than others, however, 100–150 mm (4–6 in) is the usual depth range, and about 300 mm (12 in) apart. Arums can be propagated by division of the tubers after the foliage has died down, or by seed. They like a liquid fertiliser application during the growing period.

(Arum: 'aron', Greek word for the arum. *Family:* Araceae)

Asphodeline The fragrant, hardy, spring-flowering *Asphodeline lutea* (**plate 23**), King's Spear, from the Mediterranean, is a wiry, blue-green-leaved plant, that is distinguished by its fragrance. The 25–50 mm (1–2 in), tubular, mimosa yellow, six-petalled, star-ended, green-striped flowers are borne all up the 900 mm–1.5 m (36–60 in) stem. They are followed by green-currant seeds. Propagate by dividing the roots in late winter.

(Asphodeline: modification of 'asphodelus'. *Family:* Liliaceae)

Asphodelus Do not be put off by the ugly flower spikes of the spring-blooming *A. microcarpus.* This plant, found as far apart as the Canary Islands and Asia Minor, will arise in the most poverty-striken soils. The 900 mm–1 m (36–40 in) leathery leaves are long and sword-pointed. On closer inspection it will be noted that the white, star-shaped flowers have an extremely attractive reddish pink stripe running down each petal. A collector's plant without a doubt. It has fleshy roots which can be divided late winter–early spring. *A. albus* has white flowers, sometimes with a breath of pink.

(Asphodelus: ancient Greek name. *Family:* Liliaceae)

Babiana (baboon flower) Babianas are called baboon flowers, because baboons, when hungry, will dig up the bulbs and eat them. Their 175 mm (7 in) spikes of jostling, tubed, star-shaped, 25–50 mm (1–2 in) wide, purple flowers cluster around the stem. Some red or cream-white species are mysteriously fragrant.

The 22 mm (¾ in) bulbs, which are knitted like hessian, are planted between February and early May, 100–150 mm (4–6 in) deep, 75 mm (3 in) apart, in a well-drained soil rich in well-rotted compost. They will remain until they have outstayed their welcome, when they can be dug up and separated. These plants have become pests in certain areas, so destroy any surplus bulbs. It is in large clumps that babianas provide their unqualified display, so do not be too eager to divide them. The hairy, deep green, puckered-up leaves are sword-shaped and attractively ribbed.

Babiana stricta is 300 mm (12 in) tall, and is commonly grown. *B. rubrocyanea* (syn. *B. stricta* var. *rubro-cyanea*) is smaller and has deep, brilliant, almost royal blue flowers, but as its name 'rubro' suggests has a rich, blushing crimson throat. *B. plicata* is lavender blue and deliciously fragrant. Cultivars may be listed as *Babiana* hybrids. Notable cultivars are: 'Blue Gem', dark blue flowers; 'Lady Carey', magenta-red; 'Purple Star', deep red; 'Tubergen's Blue', blue; and 'White King', white with blue anthers. (Babiana: 'babianer', Dutch for baboon. *Family:* Iridaceae)

Begonia (tuberous/rhizomatous) In Ballarat they have a festival at which large-flowered tuberous begonias (*B. tuberhybrida*) in flower are displayed. Some flowers are as large as small dinner plates. They are a big tourist attraction. The Bendigo and Melbourne Parks Departments also grow magnificent begonias. Begonia culture is attracting a large following throughout Australia.

Large-flowered tuberous begonias can be had as singles: single crisp-fringed, single frilled, single crested, single margined, single feather-edged and so on. Doubles are available as double camellia-flowered, double rose form, double ruffled, double frilled, double narcissiflora and more. And the colours? Blatant red, mouthwatering pink, sunset apricot, vivid vermillion, autumn–spring yellows, solid whites, satiny salmon pink, self-coloured with deeper coloured pencilling around the edges and so on. You can also buy tuberous begonias for planting in hanging baskets.

Tubers are available, usually August–November, and are planted in a glasshouse or similar. Once the tubers have finished flowering, and the foliage been allowed to die down naturally, those in garden beds should be lifted and carefully cleaned of dead or diseased flowers and foliage, and any soil that comes away easily. Allow them to dry off in a cool airy place. When the rest of the soil has dried it too can be gently brushed off. Those tubers in pots, once the foliage-flowers have died down naturally, can be removed from their pots and treated in the same fashion.

The tubers can be stored in shallow boxes or trays at a temperature of about 6–10°C (43–50°F). They should be inspected for fungus disease and pests, and any crown rot removed with a sharp, sterile knife, and the wound dusted with suitable fungicide. They should be inspected regularly during the rest period, with a constant check for rot. They can be covered with potting soil to prevent withering.

During August–September the tubers can be planted, right way up, into damp, gritty, granulated peat compost or John Innes compost or begonia compost, 50 mm (2 in) apart, in boxes about 100 mm (4 in) deep. Keep the tubers comfortably damp, not over-watered. Watch carefully for the tubers to sprout. When the shoots are 50–75 mm (2–3 in) tall, transfer the tubers into tight-pots (tight-potting is placing the tuber into the smallest pot that will take the roots comfortably), and grow them as pot plants, or hardened off (that is, to gradually acclimatise them to outside conditions). Plant them out when weather conditions permit.

Those to be grown as pot plants can be transferred into larger pots once they have outgrown their present pots. Take great care as the roots are brittle. Do not over-pot, because the roots do not like to be surrounded by a large area of cold, wet compost. You may have to repot once more for the more vigorous hybrids, into 200 mm (8 in) pots.

Large-flowered begonias are magnificent in garden beds. It is a moot point, but some growers have been known to keep the same tubers for up to fifty years! After that time you would need a crane to lift them! Make sure the tubers are firm, and not hollow, before you plant them. Those grown in pots will need feeding well as they are grand eaters. Use a suitable begonia, liquid fertiliser and check the label. Many growers use well-rotted cow manure in their compost.

You will soon know when begonias need feeding. When they are well fed they will proudly display their leaves. However, when they are starved they will pathetically hug their leaves close to themselves. But also check for pest and disease attack.

Tuberous begonias need ample water when they are growing well, but not waterlogged conditions. Do not spray water over the foliage as it tends to stain it, but gently direct the water into the pot or garden bed.

To obtain huge, magnificent blooms you may need to disbud or prune, to leave one or two stems, or sometimes three depending on the hybrid and age of the tuber. These stems grow tall and will need to be supported by wires or they will flop.

Tuberous begonias prefer mottled sun, as hot, direct sunlight will adversely effect the plant. Remember, do not water the plants indiscriminately. Water damages their exotic foliage and spoils the overall appearance of the display.

Favourite cultivars are: 'Lionel Richardson', peach; 'Sam Phillips', yellow; 'Flambeau', vivid orange; 'Sugar Candy', pink; 'Rosana', yellow; and many more.

Begonias which trail from hanging baskets are also available as cultivars of Begonia tuberhybrida pendula ('pendula' means to hang like a clock pendulum). 'Golden Showers' is perfect as a hanging basket specimen, but there are scarlet, cream, pink and other yellow cultivars available.

Line the hanging basket with sphagnum moss or tan bark to contain the compost. Then build it up like a layer cake. Place a layer of drainage materials such as pebbles on the bottom of the basket, then a layer of charcoal to act as a sweetener on top of the pebbles. Place the begonia compost on top of the charcoal, or an organic mixture of acid, granulated peat, ground bark and clean, sharp sand — about one-third of each. Failing this, use two-thirds acid, granulated peat to one-third sharp sand.

Insert the tubers, which have sprouted, into the hanging basket compost, and place the basket where it is to grow, bearing in mind that tuberous begonias are decimated by sharp frosts, will luxuriate in mottled sunlight, but shrivel in direct sunlight. You can buy shade-cloths for a range of plants, including tuberous begonias. These cloths allow a different percentage of light to penetrate to the plants.

Tuberous begonias in hanging baskets use up water quickly. Make sure you water them when they are in need. And don't forget to feed them!

You can take cuttings of tuberous begonias, but you will need a greenhouse in which to grow them. Place the tubers in compost, in a 100 mm (4 in) deep seed box, and start them into growth. Once the shoots appear and have produced two or more mature leaves, trace the shoot back to the mother tuber and cut off the shoot below a basal node or knuckle without damaging the mother tuber. Do not be too greedy and take too many cuttings, but leave some to grow on the mother tuber. The cuttings are inserted into a 50 per cent acid granulated peat–50 per cent clean, sharp sand compost, and grown in a warm greenhouse. Keep them shaded or the cuttings will wither.

You can also take leaf cuttings and stem cuttings. For leaf cuttings, take a healthy Rex begonia leaf, sever its main veins, and peg the leaf flat on a seed box filled with clean, sharp sand or vermiculite. Eventually roots will strike into the sand or vermiculite and new leaves will arise at the junction where the leaf was severed. These are the new plants.

Stem cutting propagation is accepted as a convenient way to increase your begonia collection true to type. Stem cuttings are taken when the plant is growing prolifically and freely. Sideshoots will be produced at leaf junctions on the main stem. Take two fingers of your hand and form a V-shape, with the index finger pointing up vertically. Growing from the flower stem junction of the V-shape you will see the sideshoot with the bud or 'eye' close against the main stem. Cut out this sideshoot when it is 100 mm (4 in) long with a sterile knife or razorblade. Take a small wedge of the main stem making sure the 'eye' is still attached to the sideshoot. Take care when taking cuttings this way as it is easy to sever the main stem. Any undue scarring of the main stem can be treated with correct fungicide. Insert the cuttings into begonia cutting compost of clean, sharp sand, granulated peat and sterilised loam and grow them in a warm, 13–18°C (55–65°F), buoyant, glasshouse atmosphere.

Tuberous begonias can also be grown from seed sown in spring. Begonia seed is extremely fine and awkward to sow. I usually mix dry, clean, white sand in the packet with the fine seed to make for easier sowing. Start by placing a drainage layer of crock at the bottom of the seed pot or pan. Cover it with a light dressing of charcoal and follow with the seed compost, firming it gently. Lightly scatter the seed over the compost and cover it with a thin layer of sand, barely covering the seed. Water the seed gently until the compost is thoroughly soaked, and let any surplus water drain away. Place the seed in a heated greenhouse to germinate, ensuring the compost does not dry out. After a few weeks the first seedlings will appear. These are followed by others, which may germinate much later. Prick out the seedlings into 50 mm (2 in) pots. They can be transferred into larger pots as they grow, but do not over-pot. You can use a compost of seven parts, by bulk, sterilised loam, two parts, by bulk, sharp sand, three parts, by bulk, granulated acid peat, and one part, by bulk, granulated cow manure. Failing this, use a 50/50 mix of sharp sand and granulated acid peat, with a suitable fertiliser added.

Begonias can be divided into tuberous, rhizomatous and fibrous clans. Of the hundreds of tuberous-rhizomatous-fibrous begonias the most popular are the so-called foliage begonias. These are grown principally for their foliage, although many have lovely flowers as well. Of the foliage beonias the Rex hybrids stand supreme.

There are as many as twenty-seven leaf shapes: angel wing, castor-bean leaf, crazy leaf, crested, oblique egg-shaped-ovate, elm leaf, fern leaf, grape leaf, heart-shaped, holly leaf, ivy leaf, lance-shaped, lettuce leaf, maple leaf, oak leaf, palm, palm-shaped, peach leaf, pond lily, rex, rex-diadema type, rose leaf, shield-peltate, spiral, star-shaped, stitched leaf, and viviparous (small leaves growing from the veins of the larger leaves, as happens on B. hispida cucullifera).

The first of the tuberous group are the 'multiflora begonias', bushy of habit and, as the name suggests, prolific flowerers, although of smaller size. They originate from Begonia davisii, an Andean species. B. davisii has been crossed with begonia species such as B. pearcei, B. micranthera and others to produce the free-flowering riot of colour we associate with the multiflora group. They are cultivated as other tuberous begonias and propagated from cuttings. They can be had as singles, semi-doubles and doubles, in white, pink, red, carmine, orange, bronze and yellow shades.

The 'Bertini' group of hybrids (B. bertini) are similar to the multiflora begonias, but are bushy plants, 300–450 mm (12–18 in) tall, carrying vermillion, single, hanging flowers of a type usually associated with the species B. boliviensis.

The origin of 'Pendula Begonias' is somewhat confused, although it has been attributed to B. boliviensis. We now know these plants as Begonia tuberhybrida pendula, and they are excellent as hanging basket specimens. The smaller rosebud and camellia-flowered are especially desirable, and the camellia-flowered 'Golden Showers' is still popular today for hanging basketwork.

To have 'Scented Tuberous Begonias' would be a great asset to any floral display, and the species B. baumannii, which has fragrance, has been crossed with tuberous hybrids to obtain tea-scented forms. Other hybridisation with other species is still continuing, particularly in the United States. (I notice that B. carolineifolia also has a delicate tea-scented rose fragrance.)

The 'American Species' are those found in Central and South America, as opposed to the Andes group mentioned previously. In this group we find the sweetly fragrant B. baumannii, B. gracilis and B. martiana, the hollyhock begonias, and B. micranthera fimbriata, which has been used to produce some exciting hybrids.

The 'Asiatic' species of begonia are found as far apart as India, China, and Japan. Perhaps the best known of these is the 'Maiden's Heart' begonia, B. evansiana, which has heart-shaped, yellow-green leaves, claret-veined below, and pale, candy-pink flowers. It is used for hybridisation with the glorious Rex begonias (plate 22) and some interesting crosses have emerged. B. picta and B. tenera are two other Asiatic species.

The 'African' species are either tuberous or semi-tuberous, and as a group they do object to being over-watered, as this causes mildew and rot to set in. It follows that their compost should be open to allow for good drainage. During cold weather some of the stems of the semi-tuberous species may die back, and this dead foliage will have to be pruned off when regrowth starts in warmer weather.

B. dregeri is famous as it is one parent of 'Gloire de Lorraine'. It is also a parent of the 'Summer Lorraine', B. 'Weltoniensis', a popular, easily propagated begonia, which has been cross-pollinated with the Rex begonias to produce miniature-leaved Rex begonias. B. sutherlandii is another fine species.

The 'Winter-flowering Begonias' are so called because in Europe they flower in winter. mostly they are hybrids from B. socotrana. Two sections have evolved, the 'English' winter-flowering, and the 'Lorraine'. Much work has been carried out hybridising B. socotrana with the Rex begonias to produce fine-foliaged winter-flowering begonias.

The 'Gloire de Lorraine' winter-flowering begonias are spoken of with reverence. Every begonia grower seems to have one or more of the countless hybrids. As pot plants they have no peer. The plant is smothered with pink or red or white flowers, with foliage that glistens with a glassy sheen. The 'Lorraines' are being cross-hybridised, and the hybrids produced are numerous and generally beautiful.

The so-called 'English Winter-Flowering' or Heimalis begonias are descendants of the Andean species B. socotrana, and to some are more lovely than the 'Lorraine' group mentioned previously. Perhaps this is because of their large flowers and colour range: rose, magenta, apricot-cream, salmon-orange, salmon pink, copper-orange and so on. They are available as singles, rosebud semi-doubles and doubles.

There are two sections of rhizomatous begonias, the Asiatic and the American, and the most famous of these. is the 'Rex' begonia originating from India. However, the American group contains some fine species, and B. manicata, and its hybrids B. m. 'Aureo maculata' and B. m. 'Verschaffelti', are still widely grown.

Apart from B. rex, the most famous of all the Asiatic species is B. masoniana, the 'Iron Cross' begonia. The name captured popular imagination as the Iron Cross was awarded for valour to German soldiers, and the leaves have dark 'iron cross' markings. This begonia is not hard to grow, and the 'goose-bump' leaves are also attractive.

If there is one plant that extracts admiration it is the *Begonia rex*, and its hundreds of cultivars (**plates 25-30**), particularly the heart-shaped, larger-leaved cultivars. They are shiny, almost metallic, in silver, green, blue-green, pink and so on. These are subdivided into the large-leaved Rex, the medium-leaved Rex, the tree Rex, the spiralled-leaved Rex, the tree-spiralled Rex, the miniature-leaved Rex, the miniature-leaved tree Rex, and the miniature-leaved tree-spiralled Rex.

We could not leave the begonia section without a brief mention of the fibrous-rooted begonias. The semperflorens group is an extremly popular, wax-flowered bedding or pot plant, particularly the F1 hybrids. The cane-stemmed group, the hairy-leaved group and the miscellaneous fibrous-rooted group are most popular in one form or another, and the reason they are not being discussed in detail is because they do not have bulbous roots.

Belamcanda *Belamcanda chinensis* is a practically unknown Chinese bulb, grown specifically for its tall, plume-like display, up to 900 mm (36 in), of 50 mm (2 in) wide, orange-red, spotted purple-brown, tubular, star-ended blossoms. The blooms are fleeting, lasting but a day, and the petals twist attractively as they fade. The flower produces 'blackberry' seed clusters, hence its common name 'blackberry lily', although the seeds look more like blackcurrants clustered around the flower stem. The sword-like foliage reaches 350–450 mm (14–18 in). The bulb requires a rich, sandy loam, and shade in the warmer parts of the country, but will survive in sheltered, open areas in the cooler areas. Plant the bulbs 50 mm (2 in) deep, 225 mm (9 in) apart. Propagate by division or seed.
(Belamcanda: Indian name. *Family*: Iridaceae)

Bellevalia A Mediterranean–Asia Minor, early-flowering, blue or white bulb similar to a cross between a hyacinth and grape hyacinth. *B. romana*, 300–450 mm (12–18 in) tall, has twenty to thirty, greenish white, 7 mm (¼ in) long flowers, flushed blue at the base, borne in reasonably compact clusters. The blue flower anthers are covered with yellow-green pollen. *B. ciliata* has 450 mm (18 in) tall, three or more, hairy, lance-pointed leaves. It has thirty to fifty, 12 mm (½ in) long, lilac-blue, greenish flowers, with purple-blue anthers, borne in a cone-like cluster. *B. latifolium*, has violet-blue flowers. The treatment for bellevalias is as for grape-hyacinths or scillas.
(Bellevalia: Pierre de Belleval, seventeenth century founder botanist. *Family*: Liliaceae)

Biarum Mediterranean bulbous plant. The species *B. davisii* has pale, yellow-green, narrow goblet-funnel, slightly hooded 'flowers' borne close to the ground, with the central true-flowering stem (spadix) rising like a snake. The leaves are small, 40 mm (1½ in) long. *B. tenuifolium* has reddish purple spathes, and 75 mm (3 in) wavy-edged leaves. Grow as pot plants in a free-draining compost.
(Biarum: ancient name. *Family*: Araceae)

Billbergia In the name *Billbergia nutans*, nutans means 'nodding', and when you see the delicate necks of *Billbergia nutans* bobbing in the breeze, you will know that they are aptly named. it is a hardy, blue-green, evergreen, bulbous plant,

which can be planted at most times. It grows well, and is a favourite, in Queensland.

It bears long necks of delicate, tubular, green, pink-striped flowers, sheathed in a translucent membrane, and the whole hangs gracefully but sadly. This billbergia is semi-hardy, thriving in well-drained soils. Propagate by dividing the roots and planting them 300 mm (12 in) apart.

B. nutans makes an excellent pot plant. It is of the same family as the pineapple, the Bromeliads.
(Billbergia: in honour of J.G. Billberg, Swedish botanist. *Family*: Bromeliaceae)

Blandfordia (New South Wales Christmas bells) *Blandfordia punicea*, a rhizmomatous-rooted plant, is restricted to certain sandy, acid soils of Tasmania, and *B. nobilis* and *B. grandiflora* are native to New South Wales, where they grow naturally, in a well-drained, sometimes moist, environment. As these conditions can be difficult to obtain in most gardens, blandfordias are often grown in shadehouses, glasshouses and the like, in acid, gritty, free-draining compost, in 225 mm (9 in) pots, where their growth conditions can be controlled.

Flowering at Christmas, these plants produce up to ten and sometimes more, scarlet, 25–40 mm (1–1¾ in), tubular bell flowers, buttercup-yellow-tipped, yellow inside, on leafy flower stems up to 900 mm (36 in) tall. The leaves are coarse-ribbed and rough to touch. *B. grandiflora* has a spike of many red bells, each tipped with yellow. *B. flammea princeps*, and the yellow *B. aurea* are also grown. They can be raised from seed or careful division of the roots.
(Blandfordia: after the Marquis of Blandford, 1766–1840 *Family*: Liliaceae)

Bletilla (Chinese ground orchid) The 'striata' in *Bletilla striata* (**plate 4**) refers to the attractively ribbed, channel pattern of the deciduous, 300–600 mm (12–24 in) leaves.

B. striata can be grown in most temperate situations. They are native to China and Japan and hardier than you would imagine an orchid to be, although they can be nipped by frost. It is therefore better to plant this orchid in semi-shade to protect it from frost.

The tuberous roots can be planted July–August after flowering, just below soil level, 300 mm (12 in) apart. They are greedy and like a soil rich in humus material such as well-rotted compost, acid peat and well-rotted farmyard manure. The heavily striated, or channel-ribbed, leaves stab through the ground in spring and are attractive as a foliage feature. So, too, is the variegated-leaf form. But it is the 300–450 mm (12–18 in) slender spikes of 25–50 mm (1–2 in) cyclamen-pink to amethyst-purple orchid flowers in late spring–early summer that set off the arrangement. *B.s. alba* is a white, pale yellow-lipped form of Chinese ground orchid, and there are cultivars available from *B. striata* and *B. s. alba*.
(Bletilla: from the resemblance to the Bletia orchid. *Family*: Orchidaceae)

Bloomeria (golden stars) The Californian *Bloomeria crocea*, having corms 12.5 mm (½ in) in diameter, is similar to the *Brodiaea* (*Triteleia*) species, and its cultural treatment is the same. It produces an original leaf, which can be 75–300 mm (3–12 in) long and 6–12 mm (¼–½ in) wide. The 15–25 mm (¾–1 in) yellow, dark-striped, six-petalled, semi-tubular, star-ended flowers are grouped on the 150–300 mm (6–12 in)

stalk in a loose, but sizable cluster, up to 150 mm (6 in) across. *B. aurea* is a golden form. Plant the corms 50–75 mm (2–3 in) apart and 50–75 mm (2–3 in) deep.
(Bloomeria: after H.B. Bloomer, United States botanist. *Family*: Liliaceae)

Bomarea (coral bells, climbing bells) Climbing bells come from a family of plants known as the Alstroemeriacae, Peruvian lily family (although some claim Amaryllidaceae, the daffodil family). Some know them as the climbing alstroemerias. *Bomarea* species is a distinct species and produces some magnificent flowers.

Bomareas originate from Central and South America, where they thrive at various high altitudes. They are tall, 1.8 m (72 in), twining, clambering plants. Some varieties have large clusters, up to fifteen, summer-flowering, 40 mm (1½ in), tubular, yellow or orange or spotted flowers. Being inhabitants of high altitudes, Bomareas suffer in the hot Australian sun, which is why they are grown in large shadehouses, or clambering along walls, fences or trees in the cooler climates of Australia.

Bomareas can be propagated by seed or by careful division of the rootstocks, planting the divisions approximately 125–200 mm (5–8 in) deep, 900 mm–1.2 m (36–48 in) apart in a loamy, organic soil.

B. caldasii (**plate 31**) is yellow-green-tipped, orange or spotted crimson; *B. wercklei*, scarlet with yellow throat; *B. edulis*, rose or yellow-green-tipped, lime-green-spotted, inner throat.
(Bomarea: in honour of Jaques Christopher Valmont de Bomare. *Family*: Alstroemeriaceae-Amaryllidaceae)

Bongardia *Bongardia chrysogonum*, a fragile, spring-flowering, tuberous plant from Western Asia, produces seven or more whorled sets of 10 mm (½ in), oval-pointed, blue-green leaves around the 45 mm (1¾ in) stem, like a series of propellors. The stems emanate from the 50 mm (2 in) wide tuber directly. The erect flower stem, which seemingly towers above the almost ground-hugging foliage, is approximately 300–600 mm (6–12 in) tall. At the top it bears small, 10 mm (½ in), mimosa yellow, star-shaped, buttercup flowers. Plant the tubers in pots, just below soil level, in free-draining sandy loam, peat compost. Keep the tubers dry in summer and not too wet in winter. A collector's plant.
(Bongardia: after A.H.G. Bongard, German botanist. *Family*: Berberidaceae-Podophyllaceae)

Boophone (Boophane) The South African-Angolan, spring-flowering *Boophone disticha*, Cape poison-sore eye, sends up, particularly after bushfires, a stocky stem from out of the bare ground, on which is borne a large 200 mm (8 in), half-circle flower cluster, similar to an agapanthus flower cluster, of a hundred or more, red-pink, starry flowers. The 300–600 mm (12–24 in), strap-like, wavy-edged, fan-shaped greyish, blue-green leaf clutch follows the flowers. The large bulbs, up to 300 mm (12 in) in diameter, which are poisonous to stock, are planted with their necks sniffing the air in a sandy, well-drained soil that receives good sunlight. Boophones stand up to drought well. They can be left in the ground, but only in areas that do not receive frosts. Leave the bulbs to colonise for many seasons as they object strongly to being shifted. In frost-prone areas the bulbs will need to be lifted and stored in a winter temperature that does not go below 13°C (55°F), which means they could be grown singly in large pots. *B. ciliaris* is a purple-flower form. (Boophane: 'boos', ox; 'phone', kill: poison ref. *Family*: Amaryllidaceae)

Bowiea *Bowiea volubilis* is an unusual South African plant in as far as the large, 100–175 mm (4–7 in) bulb produces wiry, 2–2.5 m (6–8 ft) twining stems, without leaves, on which are borne, in summer, small, 8 mm (⅓ in), greenish, star-shaped flowers. Grow in pots or in open and semi-shaded spots. Plant the bulbs in autumn in light, open soil with the bulb's neck and shoulders above ground level, 300–450 mm (12–18 in) apart. The globular, onion-like bulbs will seemingly sit on top of the ground when growing. Place stakes by each bulb, around which the emerging stem can twine.
(Bowiea: after J. Bowie, plant collector. *Family*: Liliaceae)

Bravoa (syn. *Polianthes* species) The Mexican *Bravoa geminiflora* is from the family Agavaceae, as is the Tuberose (*Polianthes tuberosa*), although both can still be found listed in the daffodil family, Amaryllidaceae. Bravoas cannot tolerate too cold a climate, and will only survive in a winter temperature of 12°C (52°F) and above. In cold climates they can be lifted, stored over the winter period, and replanted after the risk of frost is over. However, they do prefer to colonise an area for a number of years. Bravoas will flower in early summer.

They can be grown as pot plants, but would have to be repotted each year as they are vigorous growers. They need a well-drained soil and will survive in a sunny site or a partly shaded site. Fertilise with a slow-acting, low-nitrogen, granular fertiliser. The 300–700 mm (12–28 in) stems bear loose stem clusters of coral pink, yellow or orange-red, 20–25 mm (¾–1 in) long, tubular-funnel, 7 mm (¼ in) wide flowers in pairs on short stalks. The sword-shaped leaves are 300–600 mm long and 12.5 mm (½ in) wide. Propagate by division of the rootstocks late autumn–winter, or by sowing seed, which takes two years for the tubers to flower. *Bravoa graminifolia* (red) and *B. platyphylla* (pale red) are also grown.
(Bravoa: after Nicholas Bravoa, Mexican War of Independence hero; or Leonardo and Nicholas Bravoa, Mexican botanists. *Family*: Agavaceae)

Brimeura *Brimeura amethystina* (*Hyacinthus amethystinus*) is an early-flowering, Spanish, semi-alpine bulb resembling a small, delicate, 100–250 mm (4–10 in) tall, hyacinth, with up to eight or more, 10 mm (½ in) striped, single, bell flowers on a stem. The colours can vary from white to deep blue. Plant in a well-drained, sunny position, 50 mm (2 in) deep and 100 mm (4 in) apart.
(Brimeura: after Marie de Brimeur, sixteenth century gardener. *Family*: Liliaceae)

Brodiaea (*Dichelostemma*) (North American firecracker) The North American firecracker is so named because of the mass of small, 25–50 mm (1–2 in), scarlet (or purple), tubular-bell, star-shaped, early summer flowers that seem to explode at the top of strong, 300–900 mm (12–36 in), slender stems. The leaves are prostrate, few and sometimes withered at flowering time.

Most brodiaeas come from North America. *B. elegans*: clusters of violet-purple, lilac-like blooms. *B. ida-maia* produces seven to twenty or more, vivid scarlet, blue-green-yellow-tipped flowers, *B. bridgesii*: up to twenty, pale lilac to blue blooms on one cluster. *B. pulchella*, the beautiful brodiaea, has tracy, violet flowers. *B. laxa*, (Ithuriel's spear), delivers purple-pink flowers. *B. hyacinthina*: up to thirty white or purplish blue blooms in a cluster. *B. lutea* has small, yellow blooms.

You may not be able to find many of the above listed as brodiaeas, as their names may have been changed. *Triteleia* is the species name given to *B. lutea* (ixioides) and *B. laxa*. *Dichelostemma ida-maia* is the name given to *B. ida-maia* and *B. pulchella* becomes *D. pulchella*.

Brodiaeas make excellent cut flowers. Their foliage generally is grass-like, but it is the 'graceful lady' 300–900 mm (12–36 in) stems, with their heavy flower clusters that are most sought after.

The corms, which are similar to gladiolus corms, are planted in late autumn in a well-drained, preferably sandy soil. Place them 50–125 mm (2–5 in) deep in the soil, depending on corm size, 75–100 mm (3–4 in) apart. Some brodiaea species can be grown from seed. You can grow them as pot plants. In cold climates they have to be protected from frost by an eiderdown of straw or peat or similar.
(Brodiaea: in honour of James Brodie, Scots botanist. *Family*: Liliaceae)

Brunsvigia (Empress Josephine's lily) The South African *Brunsvigia josephinae* has characteristics similar to the *Amaryllis* (Belladonna lily) species, with which it is sometimes confused, although being tender it is confined to the warmer districts. Hybrids known as *Amarygia* have been produced from *B. josephinae* and *A. belladonna*.

Brunsvigia has sweetly scented, summer–early autumn flowers, with between twenty and sixty florets on a single, thick, 300–900 mm (12–36 in) flowering stem. This provides for a huge explosion cluster of cerise-orange-red, 75–85 mm (3–3½ in), narrow, tubular trumpets, 45–50 mm (1¾–2 in) wide at the mouth. *B. josephinae* has the rare distinction of having an egg-shaped bulb so large, up to 200 mm x 300 mm (8 in x 12 in), that it would barely fit in the palms of your hands. As with the Belladonna lily the 450–900 mm (18–36 in) tall, 50–200 mm (2–8 in) wide, strap-like, glaucous-blue, smooth leaves develop after the naked scape (flower stalk) has died down. The flower cluster, when dying, looks like open fingers closing. The cluster will then expand to soccer ball size, eventually falling off the plant, to tumble about in the wind.

Propagation is difficult, and is from division of offsets, rarely produced, and which take many seasons to flower. These offsets are not taken from the mother bulb until they have developed into a reasonably large size. Seed, which is scarce and rarely fertile, can take up to seven years to produce flowers. Plant offsets after the flowers and leaves have died back. Plant shallowly in a warm, sheltered, sunny site, in a well-drained organic soil, with the tip of the bulb barely peeping above the soil surface, 600 mm (24 in) apart, or singly in 300 mm (12 in) pots. Water them well during the growing season, but refrain from doing so during the dormant season
(Brunsvigia: in honour of House of Brunswick. *Family*: Amaryllidaceae)

Bulbine *Bulbine latifolium* (**plate 32**) flowers early summer and produces a mass of small, star-shaped, fluffy-anthered, mimosa yellow flowers, clustered along a 600 mm (24 in) stem. The leaves are soft green and spread on the ground like a starfish. It can be grown in a sheltered but open site and needs ample water during summer. Propagate by seed or by division of the roots after the flowers have died back and have been tidied up. Other species, such as the South African *B. caulescens*, has a 300–600 mm (12–24 in) tall flower stem, with a 300 mm (12 in) spike-cluster of yellow flowers. The leaves are 300 mm (12 in) long and soft green.
(Bulbine: 'bulbos', Greek for bulb. *Family*: Liliaceae)

Bulbinella (cat's tail) The bulbinella resembles *Kniphofia uvaria*, the yellow 'red hot poker'. It produces many lemon yellow or buttercup or orange starburst flowers along its slender spike. Unfortunately, the flowers cluster towards the top of the spike with little below which, at certain periods of its flowering life, makes the plant appear scruffy.

The South African species produce 450–600 mm (18–24 in) tall stems, topped with flower spikes 100–300 mm (4–12 in) long. The flowers are massed like corn on the cob, cramming into every millimetre of space. The star-shaped flowers have fluffy anthers that protrude way out from the petals, giving the plant an overall lacy effect.

The plants die back and disappear after flowering, and by November–December there is usually little trace of their existence. However, in February–March the attractive, fresh green, 300–600 mm (12–24 in) long foliage will appear, followed in July–September (depending on climate) by the flower spikes.

They will grow in friable, or peaty, moist soil. The bulbs can be planted out when they are dormant with the tops at ground level, 450 mm (18 in) apart. Do not leave the bulbs out of the ground to dry out. Water the plants well during the growing period, as they suffer badly from lack of water. *B. floribunda*, yellow or orange flowers, is a popular species. *B. hookeri* and *B. rossii* are New Zealand species. *Chrysobactron rossii* is a plant that in the southern hemisphere is also called bulbinella, and is related to the South African bulbinella discussed above.
(Bulbinella: diminutive of bulbine; from bulbous onion. *Family*: Liliaceae)

Bulbocodium (spring meadow saffron) *Bulbocodium vernum* is a crocus-like corm similar to the colchicum or the merendera, and is found flowering in garden beds and rock gardens early in the flowering year. Its pinky purple flower petals open and flop into a star shape as it ages. Massed in garden beds or rock gardens they make a grand splash of colour.

The 25–50 mm (1–2 in) flowers, which appear in spring, are followed by deep green, chunky, 150 mm (6 in), strap-like leaves. Plant the bulbs 50 mm (2 in) deep, 75–100 mm (3–4 in) apart in February–March, in a well-drained soil. Bulbocodium is a reluctant bulb producer, but will produce more if dug up and replanted at frequent yearly intervals. There are two, sometimes three, flowers to a bulb. *B. versicolor* has smaller flowers and more grass-like foliage.
(Bulbocodium: 'bolbos', Greek for bulb; 'koduon', head. *Family*: Liliaceae)

1. *Agapanthus orientalis* (p. 26)

2. *Amaryllis belladonna* (p. 27)

3. *Billbergia nutans* (p. 31)

4. *Bletilla striata* (p. 31)

5. *Gloxinia* cultivar (p. 59)

6. *Habranthus robustus* (p. 60)

7. *Kniphofia uvaria* 'Royal Standard' (p. 66)

8. *Leucojum vernum* (p. 67)

9. *Oxalis purpurea* (p. 75)

10. *Strelitzia reginae* (p. 80)

11. Water-lily pool (p. 74)

12. *Zantedeschia aethiopica* (p. 84)

13. *Achimenes longiflora* (p. 26)

14. *Agapanthus orientalis albus* (white form) (p. 27)

15. *Alocasia* species (p. 27)

16. *Alpinia speciosa* (p. 27)

17. *Alstroemeria* 'Ligtu hybrid' (p. 27)

18. *Amaryllis belladonna* (p. 27)

19. Anenome St Brigid (p. 28)

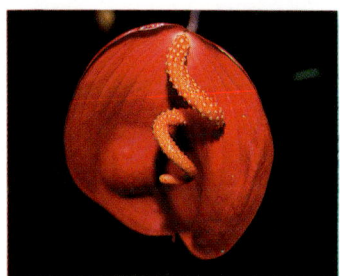

20. *Anthurium andreanum* (p. 28)

21. *Arthropodium cirrhatum* (p. 29)

22. *Arum palaestinum* (p. 29)

23. *Asphodeline lutea* (p. 29)

24. *Begonia rex* (p. 30)

25. *Begonia tuberhybrida* 'Avalanche' (p. 29)

26. *Begonia tuberhybrida* 'Amy Chart' (p. 29)

27. *Begonia tuberhybrida* 'Frilly' (p. 29)

28. *Begonia tuberhybrida* 'Guardsman' (p. 29)

29. *Begonia tuberhybrida* 'Gold Plate' (p. 29)

30. *Begonia tuberhybrida* 'Melissa' (p. 29)

31. *Bomarea caldasii* (p. 31)

32. *Bulbine latifolia* (p. 32)

33. *Caladium* cultivar (p. 49)

34. *Canna indica* scarlet cultivar (p. 49)

35. *Canna indica* purple cultivar (p. 49)

36. *Canna indica variegata* (p. 49)

37. *Canna warscewicz* (p. 49)

38. *Chasmanthe aethiopica* (p. 50)

39. *Chlidanthus fragrans* (p. 50)

40. *Clivia miniata* (p. 50)

41. *Colchicum speciosum* (p. 50)

42. *Convallaria majalis* (p. 51)

43. *Crocosmia x crocosmiiflora* (p. 52)

44. *Crocus tomasinianus* (p. 52)

45. *Crinum* 'J.C. Harvey' (p. 51)

46. *Crinum flaccidum* (p. 52)

47. *Cyclamen africanum* (p. 53)

48. *Cyclamen persicum:* drawf form (p. 52)

49. *Cyrtanthus mackenii* (p. 53)

50. *Dahlia* 'Sunset Glow' small decorative (p. 53)

51. *Dahlia* 'Bronze Delight' small formal decorative (p. 53)

52. *Dahlia* 'Edge of Gold' formal medium decorative (p. 53)

53. *Dahlia* 'Comet Red' anenome form (p. 54)

54. *Dahlia* 'Eddie Howard' exhibition cactus (p. 54)

55. *Dahlia* 'Mellisa Munchenburg' semi-cactus (p. 54)

56. *Dahlia* 'Sunset' Medium semi-cactus (p. 54)

57. Dahlia 'Seedling' nymphea form (p. 54)

58. *Dahlia* 'Koala' orchid form (p. 54)

59. *Dahlia* 'Festival' collarette form (p. 54)

60. *Dahlia* 'Margaret John' pompon form (p. 54)

61. *Dahlia* 'Coltness' bedding form (p. 54)

62. *Dahlia imperialis* (D.maxonii), tree dahlia (p. 54)

63. *Dietes bicolor* (p. 71)

64. *Eucomis cosmosus* (p. 57)

65. *Eucomis pole-evansii* (p. 57)

66. *Endymion hispanicus* (pink form) (p. 56)

67. *Endymion hispanicus* (blue and white forms) (p. 56)

68. *Freesia* 'Bergunden Mixed' (p. 57)

69. *Freesia*, scarlet hybrid (p. 57)

70. *Galtonia candicans* (p. 58)

71. *Gladiolus* 'Minniehaha' (p. 58)

72. *Gladiolus* 'Red Soft Glow' (p. 58)

73. *Gladiolus* 'George Lambert' (p. 58)

74. *Gladiolus* 'Blue Bird' (p. 58)

75. *Gladiolus* 'Jewel Song' (p. 58)

76. *Gladiolus paluster* (p. 58)

77. *Gladiolus segetum* (p. 58)

78. *Gloriosa rothschildiana* (p. 59)

79. *Hedychium gardneranum* (p. 61)

80. *Heliconia bicolor* (p. 61)

81. *Hemerocallis* 'Orientalist Peach' (p. 61)

82. *Hemerocallis* 'Honeyrock Hostess' (p. 61)

83. *Hemerocallis* 'Lemon Beauty' (p. 61)

84. *Hemerocallis* 'Sunset Orange' (p. 61)

85. *Hippeastrum vittatum* cultivar (p. 61)

86. *Hippeastrum vittatum*; pink and white cultivar (p. 61)

87. *Hippeastrum vittatum*; red cultivar (p. 61)

88. *Hippeastrum vittatum*; white-red picotee edged cultivar (p. 61)

89. *Hosta rectifolia* (p. 62)

90. *Hosta undulata* (p. 62)

91. *Hyacinth* 'Duke of Westminster' (p. 62)

92. *Hyacinth* 'L'Innocence' (p. 62)

93. *Hyacinth* 'La Victoire' (p. 62)

94. *Hyacinth* 'Princess Margaret' (p. 62)

95. *Hymenocallis* x *festalis* (p. 63)

96. *Hypoxis longifolia* (p. 63)

97. *Iris*, bearded 'Campus Flirt' (p. 64)

98. *Iris*, bearded 'Latin Lover' (p. 64)

99. *Iris*, bearded 'Desert Coral' (p. 64)

100. *Iris*, bearded 'Strathmore' (p. 64)

101. *Iris*, Dutch (p. 64)

102. *Iris tingitana* (p. 64)

103. *Ixia maculata*, deep-pink cultivar (p. 66)

104. *Lachenalia aloides* (p. 66)

105. *Lapeyrousia cruenta* (p. 66)

106. *Leucojum autumnale*

107. *Leucojum aestivum* (p. 67)

108. *Lilium auratum* (p. 67)

41

109. *Lilium candidum* (p. 68)

110. *Lilium davidii unicolor* (p. 68)

111. *Lilium longiflorum* (p. 68)

112. *Lilium martagon* 'Pendant Recurved' (p. 67)

113. *Lilium tigrinum flavum* (p. 68)

114. *Lilium* 'Enchantment' (p. 67)

115. *Lilium henryi* x *citrinum* (p. 68)

116. *Lilium* 'Wattle Bird' (p. 68)

117. *Lilium* 'Nutmegger' (p. 68)

118. *Lilium* 'Hornback's Gold (p. 68)

119. *Lilium auratum* x *japonicum* x *rubellum* (p. 68)

120. *Lilium auratum* x *platyphyllum* (p. 68)

42

121. *Lilium* 'Edward Haywood' oriental hybrid (p. 68)

124. *Lilium* 'Easter Bunny' (p. 68)

127. *Lilium* 'Apricot Aurelian' (p. 68)

130. *Lycoris radiata* (p. 70)

122. *Lilium* 'Double Atomic' cultivar (p. 68)

125. *Lilium* Asiatic hybrid (p. 68)

128. *Lilium wilsoni* seedling (p. 68)

131. *Moraea spathulata (spathacea)* (p. 71)

123. *Lilium* 'Parkman' (p. 68)

126. *Lilium* 'Aurelian Pendant' (p. 68)

129. *Liriope muscari* (p. 69)

132. *Monstera deliciosa* fruit (p. 70)

133. *Musa textilis* (p. 71)

134. *Muscari szovitsianum* (p. 71)

135. *Narcissus bulbocodium* (p. 72)

136. *Narcissus cyclamineus* (p. 71)

137. *Narcissus juncifolius* (p. 72)

138. *Narcissus pseudonarcissus* (p. 72)

139. *Narcissus triandrus albus* (p. 72)

140. *Narcissus tazette* (p. 71)

141. *Narcissus* 'Gaytime' (p. 72)

142. *Narcissus* 'Glorification' (p. 72)

143. *Narcissus* 'Maybe' (p. 72)

144. *Narcissus* 'Isobella' (p. 72)

145. *Narcissus* 'Summer Fiesta' (p. 72)

146. *Narcissus* 'Landmark' (p. 72)

147. *Narcissus* 'Double Event' (p. 72)

148. *Narcissus* 'Pia' (p. 72)

149. *Narcissus* 'Salmon Trout' (p. 72)

150. *Narcissus* 'Candide' (p. 72)

151. *Narcissus* 'Nautilus' (p. 72)

152. *Narcissus* 'Kiandra' (p. 72)

153. *Narcissus* 'Daydream' (p. 72)

154. *Nelumbo nucifera* (p. 72)

155. *Neomarica gracilis* (p. 72)

156. *Nerine flexuosa* (p. 73)

157. Orchid (*Diuris longifolia*) (p. 74)

158. Orchid (*Glossodia major*) (p. 74)

159. *Ophiopogon jaburan* (p. 74)

160. *Ornithogalum arabicum* (p. 74)

161. *Ornithogalum caudatum* (p. 74)

162. *Pancratium maritimum* (p. 75)

163. *Polygonatum multiflorum* (p. 76)

164. *Ranunculus asiaticus* (red) (p. 76)

165. *Rechsteineria cardinalis* (p. 77)

166. *Scilla hughii* (p. 78)

167. *Scilla peruviana* (blue form) (p. 78)

168. *Scilla peruviana* (white form) (p. 78)

46

169. *Sisyrhinchium iridifolia* (p. 79)

170. *Smithiantha zebrina* (p. 79)

171. *Sparaxis tricolor* (p. 79)

172. *Spathiphyllum clevelandii (S. kochii)* (p. 79)

173. *Sprekelia formosissima* (p. 79)

174. *Sternbergia lutea* (p. 79)

175. *Streptocarpus* hybrids (p. 80)

176. *Streptocarpus caulescens* (p. 80)

177. *Strelitzia reginae juncea* (p. 80)

178. *Tradescantia virginiana* (p. 81)

179. *Tritonia securigera* (p. 81)

180. *Tulbaghia fragrans* (p. 82)

181. *Tulipa sylvestris* (p. 83)

182. *Tulipa* 'Golden Supreme' (p. 82)

183. *Tulipa* 'Aristocrat' (p. 82)

184. *Tulipa* 'Campus' (p. 82)

185. *Tulipa* 'William Copeland' (p. 82)

186. *Tulipa* 'Appeldorn' (p. 82)

187. *Tulipa saxatilis* (p. 83)

188. *Vallota speciosa* (p. 84)

189. *Veltheimia viridifolia* (p. 84)

190. *Watsonia danfordiae* (p. 84)

191. *Zantedeschia elliotana* (p. 85)

192. *Zephyranthes candida* (p. 85)

Caladium Tropical South American 'Fancy-leaved' caladiums **(plate 33)** are in the gold-medal class of the foliage plants, with long-lasting, heart-shaped or spear-shaped leaves as large as your hand and wrist. These leaves range in colour from copper-silver to splashed green and gold; breathless white on delicious green to green and gold; blood red and green or pastel-pink to green and silver.

Being tropical, caladiums like heat, and a growing temperature of 21°C (70°) would be reasonable. They are grown in the heated greenhouse or a warm part of the home in the southern states. In warmer climates they can be moved, still in their pots, to a bush-house or shadehouse, when all fear of frost and other inclement weather is over.

The tubers are planted in pots during the dormant period. They usually become available in nurseries in September–November. Use a sharp sand-peat compost as a potting mix, and allocate one to three tubers to a 125 mm–175 mm (5–7 in) pot. Crock the bottom of the pot to ensure that surplus moisture drains away. Crocking is the placing of pebbles or broken pieces of clay pot convex over the pot drainage hole. Half fill the pot with compost. Plant the tuber and nearly fill up the pot with the rest of the compost. Leave a 38 mm (1½ in) gap at the top of the pot to prevent soil and water being washed over the side of the pot when you water.

Water sparingly until the leaves are growing well and then increase the quantity. Caladiums need feeding only when the leaves are growing well. Use a weak liquid fertiliser. You can buy fertilisers specially compounded for plants such as caladiums. Place the plants in a light, airy, but warm situation, shaded from the direct rays of the sun. Remove the leaves carefully when they die. Store caladiums when they are dormant in a 13–15.5°C (55–60°F) temperature.
(Caladium: from the Indian 'kelady'. *Family:* Araceae)

Caladena (see *Orchids*)

Calanthe (scrub lily) (see Orchids)

Calla (See *Zantedeschia*)

Calochortus (mariposa lily or tulip or butterfly tulip) 'Mariposa' is Spanish for butterfly. mariposa lilies hover above the grass-like foliage, like a swarm of butterflies flitting across a meadow. To some they look more like upturned, split bells. Mariposa lilies are native to western North America, particularly the warm Pacific coast. They like perfect drainage, as waterlogging rots the bulbs.

These lily–tulip relatives can be planted in a light, shaded site or deciduous woodland. They flower spring–summer and produce fancy, goblet-shaped blooms in their emerging stage, in shades of white, yellow, lavender, pink and red. The buds are long and pointed, forming an interesting feature. Once the flowers open they produce 50 mm–125 mm (25 in) wings, depending on species, and expose spotted or blotched throats, which adds to their attraction. There are some species whose flowers hang down, and these are known as the lantern types.

Popular species are C. *venustus*, white, yellow, red or purple flowers, blotched inside brown-red; C. *albus*, the white, nodding lantern; C. *amabilis*,

the golden lantern, buttercup yellow; C. *pulchellus*, the golden lantern, lemon yellow; and C. *caeruleus (coeruleus)*, the blue (more white than blue) mariposa. There are others. They can be increased from ground-bulb offsets, or aerial axil-leaf bulbs, according to species, and will grow from seed, which can take four or five years to flower. Plant the bulbs in autumn in an open soil, with clean, sharp sand added, 50–90 mm (2–3¾ in) deep, depending on whether the soil is heavy or sandy, 150 mm (6 in) apart. The young offsets may not flower for about two years, but be patient as the butterfly flowers, borne on 250–900 mm (10–36 in) stems, are worth waiting for.
(Calochortus: 'kalos', beautiful; 'chortus', grass — grass foliage. *Family:* Liliaceae)

Caloscordum (Nothoscordum) *Caloscordum neriniflorum*, from eastern Siberia and China, is an onion-like-foliaged, and miniature nerine-flowered plant, 150 mm (6 in) tall, with an upright cluster of up to twenty, spidery fingers on each onion-like shoot, on which are borne 10 mm (½ in) pink, star-shaped flowers. An early flowering bulb. Grow in pots in a free-draining compost. A collector's plant.
(Caloscordum: 'kalos', beautiful; 'scordum', garlic. *Family:* Liliaceae)

Calostemma (Australian bell) The most important species is C. *purpureum*, which produces clusters of rose pink-purple, hanging, funnel, bell-flowers. It flowers late summer–autumn. They divide easily from clumps, and the seed they produce is viable and can be readily grown.

C. *purpureum* is a sun-lover from the interior, and as such likes a warm climate, sheltered position and well-drained soil. The bulbs are planted May–early June and, depending on bulb size, from 75–150 mm (3–6 in) deep (the big bulbs being planted the deepest), and 150 mm (6 in) apart depending on clump size.
(Calostemma: 'kalos', beautiful; 'stemma', crown. *Family:* Amaryllidaceae)

Camassia (quamash; camass) Camass lilies in their natural state grow on the Pacific coast of California, northwest America and British Columbia, and were of great economic importance to the North American Indians. They are naturalised in American gardens. The flowers, which are borne on racemes (spikes) up to 600 mm (24 in) tall, depending on species, are slim, six-petalled, star-shaped, with well-extended centre parts. The colours can range from white to cream to blue to lavender to purple, indeed, C. *quamash (esculenta)* produces long scapes (leafless stems) of variable white, blue or purple flowers on 450–600 mm (18–24 in) stems. C. *leichtlinii* is a 1.3 m (52 in) tall species of camassia, which sends up flower scapes (stems) that can reach a metre or more in height, and throws blue, blue-white or purple flowers.

Plant camassia bulbs in April–May, 75–120 mm (3–4½ in) deep, depending on bulb size and climate, 150 mm (6 in) apart, in a rich, well-drained soil. They like moisture during dry spells. They bear their flowers in late September–October depending on climate. Transplant the bulbs when they are dormant. The smaller species produce many small bulbs, but the larger species, generally, produce only a few large bulbs. Species can be grown easily from ripened seed, sown in

autumn, which can take up to five years to flower.
(Camassia: American Indian name. *Family:* Liliaceae)

Canna (Indian shot) Why is Indian shot so called? There is hardly a place in temperate Australia where summer-flowering cannas **(plates 34-7)** will not grow, provided they are planted in a rich soil and receive adequate water. It is an exotic plant, originating from Central America and the West Indies. A massed bed of turkey-head, scarlet cannas reminds one of the piracy and pageantry of the old seventeenth century West Indies, hot-looking and full of bright colour.

There are at least fifty-five species and hundreds of cultivars to choose from, both dwarf and tall, 500–1800 mm (20–72 in), with a range of differing foliage colours. My favourite is the one that produces deep orange flowers and variegated leaves. The colour range of canna flower spikes is fabulous, running through all the hues and shades of red and yellow.

They are best planted just below ground level, 750–900 mm (30–36 in) apart, massed in beds of their own, where they will soon spread and need to be divided. They require a top dressing each year of well-rotted farmyard manure. Failing this, use a complete bulb fertiliser. Plant out the tuberous roots August–September.

The flower stalks should be cut down once the cannas have finished flowering. The tuberous roots can be left in the ground. In the colder parts the roots can be covered with a bedspread of straw. In hard, frost-prone areas of Australia and New Zealand they should be lifted and stored, similar to the way one would treat dahlias in like conditions.

Cannas need abundant water during dry spells. They also like a rich, moisture-retentive soil, but not waterlogged. The tuberous roots can be planted singly in large pots as a decoration. However, should you grow them in pots, use those cannas with attractive foliage, such as the bronze and variegated foliage cultivars. Flowers will appear continually over a long period.

You can grow cannas from seed, but you would get only poor-looking species unless you are cross-breeding your own varieties (cultivars). The seed is hard and black like blunderbuss shot and that, incidentally, is where the plant gets its common name. The revolutionaries fighting the British, so the story goes, were so poor that they could not afford lead shot, so they used the hard-coated canna seed. Chip the hard-coated seed, or soak it in warm water for twenty-four hours before planting.

Seed can be sown spring–summer in the warmer states, outside in a seedbed that has been broken down into a fine tilth; and autumn–spring in the cooler states, in pots in a sheltered site. Cover the seed with 12 mm (½ in) of fine soil. Water the seed regularly if need be. When the seedlings are large enough to handle, prick them out into seedbeds 150–200 mm (6–8 in) apart. Transfer them to their permanent quarters when they have matured.

Do not let cannas run to seed, unless you feel it necessary or are cross-breeding them, as it drags much goodness from the tuberous roots.
(Canna: possibly Latin or Greek for cane-like stems. *Family:* Cannaceae)

Cardiocrinum *Cardiocrinum giganteum*, the giant heart-lily, brazenly displays large, 150–

300 mm (6–12 in) trumpet flowers, up to twenty to a stem, white-cream and slightly down-pointing, with purple-red blotched throats, in December–January.

It has found its niche in the temperate mountain areas of Australia. It is exotically perfumed. Its 500 mm (20 in) long, deep green, heart-shaped leaves make it a curiosity, and it can grow as tall as the eaves of your house — 2–4 m (80–160 in).

Cardiocrinum g. var. *yunnanense* is shorter with red-bronze leaves. *C. cathaycanum*, greenish white trumpets, and *C. cordatum*, white, purple-blotched trumpets, are smaller species. The giant heart-lily prefers a cool, shady, rich organic soil. Plant the 'roof shingle' bulbs early spring, with their tips at ground level, 900 mm–1.2 m (36–48 in) in a sheltered site.

(Cardiocrinum: 'Cardio', heart; 'crinum', lily. *Family:* Liliaceae)

Chasmanthe *Chasmanthe aethiopica* **(plate 38)** is a frost-tender plant similar to the gladiolus in growth habit. The 40 mm (1½ in) long, narrow, tubular, mimosa yellow or burnt orange flowers are borne either side of the 900 mm–1.5 m (36–60 in) flower stems like a long, fishbone plume. The 600 mm (24 in) tall, light green leaves are 50 mm (2 in) wide, pointed and arranged in a fan. As a woodland plant feature in the Adelaide Botanic Gardens they have adapted successfully. Plant the corms just below ground level 300 mm (12 in) apart in early autumn. This South African plant, and the nine other species, needs to be contained, and any surplus bulbs destroyed.

(Chasmanthe: Greek 'chasme', gaping; 'anthe', flower. *Family:* Iridaceae)

Chiloglottis (see *Orchids*)

Chionodoxa (glory of the snow) The name chionodoxa comes from Greek and means snow-glory. It is so named because on the mountains of its native habitat in southern Europe, Turkey and India, it is found emerging from melting snow. Chionodoxas are grown in Victoria and the cooler parts of New South Wales, but do not thrive in hot, dry conditions. They look best in a rock garden or massed under deciduous woodland trees. The bulbs are planted in autumn 75–150 mm (3–6 in) deep, depending on bulb size, in a well-drained, gritty soil, in clumps of six, allowing 75–100 mm (3–4 in) between bulbs.

The flowers, three on a 100–150 mm (4–6 in) stem, appear in August–October depending on climate. Eventually the flowers and foliage will die down. Leave the bulbs in the ground to increase. Cover them with a thin, 25 mm (1 in) layer of compost mulch, as this will help feed them. Keep down weeds around the plants. Chionodoxas can be used also as pot plants.

The species most readily available is *C. luciliae*, with its 15 mm (½ in) wide, star-shaped, mauve-sky blue, white-centred flowers, and its varieties, *C. l. alba* (white), *C. l. rosea* (pink), *C. l. gigantea* (which has fewer, but larger, 20 mm (¾ in) wide, mauve-blue blooms). *C. sardensis* is a deep blue-mauve. *C. siehei* is a proliferous, mauve-blue, white-centred, large-flowered plant. *C. albescens* is white-flowering but sometimes has a breath of pink or blue.

(Chionodoxa: 'chion', snow; 'doxa', glory) *Family:* Liliaceae)

Chionoscilla x *C. allenii* is a hybrid between *Chionodoxa luciliae* and *Scilla biflora*, hence the name. It has characteristics of both parents, but looks more like *S. biflora*. The 'x' in front of the botanical names signifies that it is a natural hybrid. It has charming, lilac mauve-blue, 20 mm (¾ in), star-shaped flowers, petals fused at the base, borne on 100 mm (4 in) stems. Plant the bulbs in autumn 50–100 mm (2–4 in) deep, 75–100 mm (3–4 in) apart, in groups of six to twelve or more, depending on the size of the area available.

Chlidanthus (sea daffodil) *Chlidanthus fragrans* **(plate 39),** sea daffodil or Peruvian daffodil, has six or more shiny, penetratingly fragrant, 75 mm (3 in) long, six-pointed-petalled, canary yellow, flower trumpets, borne on one 250 mm (10 in), naked scape (flower stem). The fresh green, daffodil-like foliage appears after the flowers, and dies down in winter.

This late spring-flowering, South American plant likes a sunny, rich, well-drained soil, and ample water when it is growing well and until the foliage dies down. The bulbs can be planted during the dormant season, similar to planting daffodils themselves: that is 125–150 mm (5–6 in) deep, 225 mm (9 in) apart. Propagation is by separating the bulbs during the dormant season. They can be grown as pot plants in a cool greenhouse in frost-prone areas.

(Chlidanthus: 'chlidanos', delicate; 'anthos', flower. *Family:* Amaryllidaceae)

Clivia (Kaffir lily) The drought-resistant South African Kaffir lilies grow in the most unlikely places, such as under trees, in shady positions, where their 300–450 mm (12–18 in), strap-like, glossy evergreen leaves seem to multiply and multiply, suddenly producing an eruption of clear, scarlet-orange blooms to lighten up the gloom of those dark places.

They like a shaded site, a rich, organic, well-drained soil and ample water when they are flowering and during dry summer spells. They also like an occasional drink during the winter, but this is usually supplied by rain. Plant them just below ground level 300–450 mm (12–18 in) apart.

Clivias will grow successfully in a shadehouse in pots or tubs, in free-draining, rich organic compost. Water them when they are growing and flowering, and during dry summer spells. Give them an occasional drink, if necessary, during the winter.

Clivias grow well in most temperate areas, but there are places that are too cold, as heavy frost will scorch the leaves. In frost-prone areas they can be grown in a shadehouse. In severe frost areas, they can be grown shaded in pots in a greenhouse.

Outside you can topdress the soil around the roots, not over the leaves, with well-rotted farmyard manure. Pot plants can be fed with liquid fertiliser. Stock is increased by careful division of the thick rootstock clumps from late summer on, after flowering.

Kaffir lilies will, depending on local climate, come into flower in late winter–spring. *Clivia miniata* **(plate 40)** which is most commonly grown, produces a fountain-cluster of ten to twelve, 50–75 mm (2–3 in), funnel-shape flowers on one 450–600 mm (18–24 in) stem. The colour is sun-kissed, apricot orange, the colour one gets when the sun dissolves in the west, with lighter yellow-orange throat.

There are hybrids available, and a useful one is *Clivia* x 'Cyrtanthiflora', a clearer orange than *C. miniata*. *Clivia nobilis*, with its cluster of thirty or more, orange-yellow, green-tipped flowers, is also grown, but it has proved most sensitive to frost, although its drooping, tubular, bell-ended flowers are a joy. There are also variegated-leaf hybrids. The Kaffir lily is popular as a cut flower, because of its iron stalk, explosion of flowers, and attractive orange seed pods.

(Clivia: after Lady Clive, Duchess of Northumberland. *Family:* Amaryllidaceae)

Codonopsis *C. convolvulacea* is a rare, high-altitude, summer-flowering Tibetan–Chinese plant that bears solitary, lilac-blue, five oval-pointed-petalled, star-shaped flowers. The petals are split nearly to the centre and are borne on 1500 mm (60 in), wiry, twining, climbing stems emanating from underground tubers. It has 50 mm (2 in) long, alternate, deep green, narrow-oval, lance-like, pointed leaves. It can be grown in cool areas. Plant the longish, tuberous roots in pots in a well-drained, peat-sand compost, or amid low-growing shrubs, provided the soil is a well-drained, sandy soil, rich in leaf mould or peat. *C. c.* var. *forrestii* is a more robust, larger-leaved plant with a pronounced purple-blue ring at the centre of each flower. The Himalayan *C. ovata*, is popular. It is 150–300 mm (6–12 in) tall. The lower part of the stem lies on the ground, with the upper part standing erect. It is hairy and has 18 mm (¾ in) long leaves. The light blue flowers are 25–37 mm (1–1½ in) long, bell-shaped, and are borne on 75–150 mm (3–6 in) stems.

(Codonopsis: Greek 'codon', bell; 'opsis', resembles. *Family:* Campanulaceae)

Colchicum (autumn crocus; meadow saffron) The 100 mm (4 in) upright flowers of the colchicum, autumn crocus, are slim, tulip-cup, crocus-like, pink or white, and appear with no stalk directly from the corm, before the foliage in autumn. The colchicum is not a true crocus but belongs to the lily family. Technically, the colchicum has a superior ovary borne above the petals and has six stamens. The crocus has three stamens. The leaves are significantly different.

The colchicum is a native of Asia Minor and Europe and will not tolerate the more tropical parts of Australia, but will thrive in the cool, neutral soils of Victoria, Tasmania and New Zealand. it will also grow in cooler parts of New South Wales and South Australia.

There are forty or more colchicum species. The 30 mm (1¼ in) diameter corm can be planted late summer, the neck of the corm being 50 mm (2 in) below soil level, each corm 100–225 mm (4–9 in) apart; some like planting the corms closer, others wider. The leaves will appear in spring and die down during summer. Propagation is by separating the offsets from the mother corm, once the leaves have died down. Colchicums, although they are not over-fussy, like a deep, rich, well-drained soil, not too exposed to the hot sun. They are grown in rock gardens and borders, in a spot where the dying leaves will not jar.

Colchicum autumnale is a pale rose or white species, commonly known as the autumn crocus; *C. speciosum* **(plate 41)** is sweetly scented with rosy pink petals, white at the base; *C. speciosum*

'Album' is as white as snow; C. *agrippinum*, an old favourite, is light pink, lilac-purple.

There are many beautiful hybrids available; such as C. 'Autumn Queen', with its chequered, white throat against a rosy purple background; C. 'Lilac Wonder', an exquisite, self-coloured violet; C. 'The Giant', rosy lilac, with large flowers; C. 'Conquest', rich, rose pink.

Colchicum luteum is the canary yellow, spring-flowering, meadow saffron. There are also pink or white early-flowering colchicums, and some have well-developed leaves at flowering time. C. *crocifolium* is pink; C. *fasciculare* is white or pink; C. *libanoticum* is pink; C. *steveni* is yellow.

A close study of the colchicum's petals reveals a veining reminiscent of marble, that is its great attraction. The large, white (album) forms are dazzling in size and marbling.

The autumn crocus leaves will follow after the flowers have died down, and it is during this period that all the goodness is produced for the next year's flowers. C. *autumnale* has leaves 150–250 mm (6–10 in) long and 25–37 mm (1–1½ in) wide. C. *speciosum* are broad-oblong, 300–450 mm (12–15 in) long and 40–80 mm (1½–3 in) wide. The leaves develop in spring. They may look untidy, but if the brown, tunic-coated corms are planted in a sensible place in the beginning, the leaves will not be an eyesore in the rock garden or border.

(Colchicum: from Colchis in Asia Minor. *Family:* Liliaceae)

Colocasia (elephant's ears; taro root) Colocasia is called elephant's ears because the leaves, when fully grown, represent elephant's ears, although to some they look like giant, 600 mm (24 in) arrowheads.

Tropical Asian colocasias can be grown in most warm parts of Australia, in a shady site. However, they are frost-tender and in frost-prone areas should be accorded the same treatment as caladiums. They are grown as far south as South Australia. Usually they are planted amid tree shade or in a shadehouse which gives protection. Unfortunately, one sees potted colocasias stuck out in the hot sun, which invariably turns the leaves a sickly yellow. The leaves are arrow- to heart-shaped and the foliage will grow 1–2 m (40–80 in) high. Being fleshy-rooted plants they require much water during the growing season and can be grown in a warm, damp, but not waterlogged soil.

The 150–300 mm (6–12 in) tall, pale green, arum-like flower (spathe), which appears late October–December, nearly envelops the centre, 'thick pencil', yellow spadix (true flower). The spadix is penetratingly fragrant and eventually produces clusters of eye-catching, 'redcurrant' seeds. But it is the swamp-like appearance of the leaves you seek. Propagate by carefully dividing the parent plant's fleshy rootstock in spring. Plant the sections 1.75 m (6 ft) apart. They should be left to colonise the area for years in warm, frost-free areas. Elsewhere, they should be treated like caladiums, or indeed, dahlias, that is stored in a frost-free place and planted out when the risk from frost is over. *Colocasia antiquorum* and C. *a.* var. *esculenta* are favourites, but there are others with leaf marking variations.

(Colocasia: from the Arabic 'kolkas'. *Family:* Araceae)

Commelina (blue spiderwort) Coming from the same family as the Wandering Jew plant (*Zebrina*

pendula), and the *Tradescantia* species, *Commelina coelestis*, the Mexican blue spiderwort looks the odd one out. Its 25–30 mm (1–1.25 in), butterfly flowers are a startling, unnaturally bright blue, and are borne on 450–600 mm (18–24 in) stems, arising from tuberous rootstocks, from spring to autumn. C.c. *alba* is a white form; C.c. *variegata* is blue and white. Divide rootstocks, and plant out late winter–early spring, 50–100 mm (2–4 in) deep, in a light, well-drained soil, where they can be easily controlled, as they do spread. C. *coelestis* makes a useful pot plant.

(Commelina: after K & J Commelin, seventeenth century botanists. *Family:* Commelinaceae)

Convallaria (lily of the valley) The small, nodding, 7–8 mm (¼–⅓ in) long, white bell-flowers of the temperate European lily of the valley, when picked, bunched and placed in the home, will dominate with their choice, sweet perfume.

Convallaria majalis (plate 42) is propagated by rhizome roots. It is a vigorous little plant and can get out of control if care is not taken to control it. Convallaris can be naturalised in cool area, semi-shaded garden sites, in a deeply dug, firmed soil enriched with well-rotted manure.

Plant out plump, firm crowns, or pips as the roots are known to old gardeners, March–April, with the top of the root 25 mm (1 in) below soil surface, 50–75 mm (2–3 in) apart. Eventually the area will be overcrowded with roots, and then the plants can be dug up, separated and replanted in a soil enriched with well-rotted farmyard manure.

The flowers will appear in late winter–early spring, on one side of the 100–150 mm (4–6 in) flowering raceme (stem), from behind longish, parallel-veined, plastic-looking leaves, and last for two or three weeks. Cut off the old stalks once the flowers have finished, but keep the plant watered for the rest of its growing period, as the leaves are producing food for the creeping rootstocks. Topdress the beds autumn or winter, with well-rotted cow, horse or sheep manure and a bulb fertiliser, added as the manufacturer instructs.

There are exciting hybrids available. The large-flowering C. *majalis* var. *major* and C. *majalis* var. *rosea* are becoming increasingly popular, although the bell-flowers tend to lose their cleanliness of shape. There is also a variegated-leaf form. Lily of the valley makes a superb pot plant when in flower, and is easily grown this way, planted in John Innes No. 1 Compost or similar.

(Convallaria: 'convallis', a valley, its habitat. *Family:* Liliaceae)

Cooperia (Zephyranthes) (giant prairie lily) The North American *Cooperia pedunculata* (*Zephyranthes drummondii*) is a summer-flowering plant. It carries on each single 225–450 mm (9–18 in) stalk, an up-facing, usually night-flowering, tubular-stemmed, solitary bloom, ranging in colour from white to primrose with a blush of red. The flower opens into a 75–100 mm (3–4 in) long, six-petalled, 37–50 mm (1½–2 in) wide, goblet. The foliage is occasional, grass-like and narrow. Increase stock by planting bulb offsets 100 mm (4 in) deep, 225 mm (9 in) apart, in a sheltered, open site. (I have seen this listed under *Zephyranthes*).

(Cooperia: after Joseph Cooper, an old gardener. *Family:* Amaryllidaceae)

Corydalis Corydalis species, natives of northern hemisphere temperate regions, produce early-

flowering 10–20 mm (½–1 in) long, curved-tubular, drinking-horn, white, pink, purple-violet or blue drooping flowers that, at the mouth, resemble snapdragons (antirrhinums). C. *solida* (*bulbosa*) has pink or purple-violet flowers; C. *emanueli* has sky blue flowers; C. *glaucescens* white with a breath of rose; C. *macrocentra*, canary yellow, with pale orange flower lips. The tuber-like root sends up 125 mm (5 in), and taller, spikes of flowers, and pale or bluish green leaves that are reminiscent of buttercup or split clover leaves.

The *Corydalis* species, mainly, are natives of southern and central Europe, central Asia and North America, and are found naturalised in stony hillsides and deciduous woodlands. They can be planted in shady borders 50 mm (2 in) below soil level, 75 mm (3 in) apart. However, they are best grown in pots, in free-draining compost in a shady place, as a curiosity or collector's plant. Some place the corydalis in the 'poppy' family, others in the 'fumitory' family.

(Corydalis: 'korydalis', flower spur resembles a larkspur. *Family:* Papaveraceae/Fumariaceae)

Corysanthes (see *Orchids*)

Cotyledon (C. *lutea*: see *Umbilicus*)

x **Crinodonna** (syn. *Amarcrinum*) A intergeneric hybrid of Crinum *moorei* and *Amaryllis belladonna*. Collector's plant.

Crinum This beautiful, bulbous, funnel-shaped, exploded flower-end genus of plants is found growing naturally in places as far apart as Australia, Asia, Africa and South America.

Crinums thrive in a well-drained soil which has been enriched with well-rotted cow or horse manure. However, C. *bulbispermum* can be grown in a moist soil, as well as a well-drained soil, provided it is watered when growing well. Crinums also like an annual topdressing of well-rotted manure, applied during the winter dormant period, with a bulb fertiliser added at the manufacturer's recommended rates. They flower early summer onwards and, if left undisturbed in clumps for a few years, will reward by producing towering clutches of fragrant, tubular, star-mouth, bell-like blooms each summer.

Plant the large 135 mm (5½ in) bulbs 50 mm (2 in) below the soil surface, 225 mm (9 in) apart, in the warmer parts of Australia, during autumn–spring in an open, sunny site. In the cooler parts of the country have the bulb noses just sniffing the air. Crinums can be increased by dividing up the clump. Crinum foliage is fleshily attractive, resembling the leaves of clivias or agapanthus. Crinums are mostly evergreen, but there are deciduous species.

Clean old leaf litter and debris carefully from the clumps. Watch for snail and slug attack and other pests, as they consider evergreen crinum leaves a delicacy. Treat with slug and snail bait. Be careful not to damage the necks of the bulbs when working around the bulb area, as contained within the necks are next year's flowers.

C. x *powellii* (Cape lily) is a hybrid between the South African C. *moorei* and C. *bulbispermum*, and is the most popular of the crinums, being sweetly fragrant, with cultivars available in soft pink or white or red flowers. The large, up to 100 mm (4 in), trumpets are borne well above the evergreen foliage on tall, 600 mm–1.5 m (24–60 in) stalks, which arise from 150–200 mm (6–

8 in) diameter bulbs. C. *moorei* is also popular as it will grow in shade. It has white trumpets flushed with red inside, and splashed green on the outside, and can bear up to ten flowers on one spike. C. *bulbispermum* (C. *longifolium*) is the giant of the family, reaching up to 1.2 m (48 in) in height. It is hardy and can be planted in awkward places, particularly damp spots. The 100–125 mm (4–5 in) pink or white blooms gladden dull days.

C. *asiaticum* the summer-flowering, asiatic poison bulb, bears up to twenty white flowers in an umbel-cluster on 1.2 m (48 in) stems, each flower being 75–100 mm (3–4 in) long. C. *a. procerum* has larger flowers, tinged red. These plants are semi-evergreen and subtropical, but will withstand temperate winters if planted in a sheltered place. The popular C. *caribaeum* has three to four blooms, 75–100 mm (3–4 in) long, on one stem. C. *erubescens* produces up to six fragrant flowers on one stem, the flowers being white inside the trumpet and purple-pink outside.

C. *campanulatum* is the small-trumpeted water-crinum, which can be planted, in pots, in 300 mm (12 in) deep ponds, or as bulbs in wet or damp places. The pots can be lifted during the dormant season and stored in a sheltered place, but they must not be allowed to dry out; just keep them pleasantly moist. Six to fifteen white or blush pink, red-striped trumpets, 50–125 mm (2–5 in) long, are borne on each 450–600 mm (18–24 in) flower stem.

C. *flaccidum* (plate 46) is our native crinum, the Murray–Darling lily, so named because it is found on the upper reaches of these rivers; it was actually discovered in the Macquarie Ranges. The 100 mm (4 in) wide, usually white (they can vary from white to pale yellow), fragrant, six-point, trumpet-like blooms arise out of deep green, thick, strap-like leaves. Eight to twenty may be on one solid, 450 mm (18 in) stalk.

(Crinum: 'krinon', a lily. *Family:* Amaryllidaceae)

Crocosmia (x *Crocosmia crocosmiiflora*; *Tritonia* x *crocosmiflora*) *Crocosmia* x *crocosmiiflora* (x *Crocosmia crocosmiiflora*) (plate 43), once known by the beautiful name of Montbretia, is a South African corm that will grow readily in Australian gardens. Plant crocosmia corms only in those areas where they can be contained easily.

The graceful-necked crocosmias will flower summer to autumn, arising from thin, bayonet, 400–600 mm (16–24 in), pleated-fan foliage. The flower colour can vary between hybrids from butter yellow to burnt orange. The miniature gladiolus-type flowers, a score or more, zig-zag like butterflies up the 1 m (40 in) stems.

Plant the corms late autumn–winter, after flowering, in a well-drained soil, 100 mm (4 in) deep and 150 mm (6 in) apart, in clumps so that they flower in groups. Leave to multiply. Apply a low-nitrogen bulb fertiliser as the manufacturer instructs. Divide the clumps when they stray outside their alloted area and destroy any surplus corms.

The original *Crocosmia* (*Tritonia*) *aurea* cross-pollinated naturally with *Crocosmia* (*Tritonia*) *pottsii* to produce *Crocosmia* x *crocosmiiflora*. This resulted in 25–40 mm (1–1¾ in) long, tubular flowers, ending with six short, star-shaped petals, 35 mm (1½ in) wide. Cultivars are: 'Vesuvius', reddish orange; 'His Majesty', orange-red variegations; 'Lady Oxford', light yellow; 'Lady Wilson', buttery yellow; 'James Cory', red blotched-brown; 'Comet', bronzey red; there are many more.

(Crocosmia: 'crocus', saffron; 'osme', smell. Dried flowers immersed in water smell of saffron. *Family:* Iridaceae)

Crocus There are more than eighty species of crocus growing naturally in Europe, the Middle East and Near East.

The variable Australian climate is not generally suitable for crocus growing. However, they will grow in the cooler parts of Victoria, New South Wales and Tasmania, and I have seen them growing in the cooler parts of South Australia. People living in Sydney, Queensland, Western Australia and South Australia may not get the effect they would like to achieve. Try planting the corms 125 mm (5 in) deep, or you may have to buy imported corms each year.

Crocus corms are planted February–March in a cool, shady position, in a light, well-drained soil. They like a reasonably moist but not waterlogged, soil. Plant in clumps of 12 to 18 corms, 75 mm (3 in) apart and 75–100 mm (3–4 in) deep. Leave them undisturbed to colonise the area. Once the clumps have outgrown their alloted area, and that will take time, dig up the corms, separate them and replant the largest and healthiest in the best spot. You can plant the smaller corms elsewhere in the garden.

The slender, spiky, grass-like foliage, with its silver midrib, will appear first, followed by the narrow, orange, yellow, purple or striped, flowers, 100–150 mm (4–6 in) tall. Let the foliage die down naturally after flowering. Crocuses can be grown naturalised in deciduous woodland and in lawns. However, grasses such as buffalo or kikuyu will smother bulbs.

Crocuses can be grown as window box specimens provided the site is not too hot. They can also be grown in bowls or pots, which can be moved into cool spots.

There are many species of crocus: C. *vernus* is usually faint lilac pink, but other colours, from snow white and lilac to royal purple, are available. A subspecies, C. *v.* ssp. *vernus*, is an ancestor of the purple Dutch crocus. C. *angustifolius* (C. *susianus*), 'Cloth of Gold' crocus, is dark canary yellow. C. *biflorus*, 'Cloth of SIlver', usually lilac-pink, is a variable species, available also in lilac-blue and white. C. *tomasinianus* (plate 44) has pale lilac-purple, white-throated, star-shaped flowers. C. *chrysanthus*, a famous species, is canary yellow, with purple outside feathering and black centre stamens. The cultivars of C. *chrysanthus* and C. *biflorus*, such as C. *c.* 'E.A. Bowles' which has rich egg-yolk inside, and tracy, purple feathering outside, are superb. There are many other cultivars ranging in colour from white, lilac and buff to pale and deep yellow, mostly with the characteristic purple streaking on the outside petals.

The giant Dutch hybrids are the 'stars' of the crocus world, which is understandable if one considers they have hybrid vigour, produce large flowers, and are available in a colour range of white, yellow, purple and striped. Plant them 75–100 mm (3–4 in) deep and 75 mm (3 in) apart in clusters of twelve or eighteen massed in groups of, say, a hundred corms.

The true, autumn-flowering crocus is a delight and there are many species to choose from: C. *sativus*, lilac-violet saffron-crocus, so called because saffron was extracted from its central organs. C. *longiflorus* has fragrant, spicy, pale rose or white flowers. C. *kotschyanus* has variable subspecies with flowers ranging in colour from

white to blush pink and striped pink.
(Crocus: ancient Greek for saffron. *Family:* Iridaceae)

Curcuma (queen lily) *Curcuma petiolata* is a 'ginger family' plant from India, which needs the same treatment as other ginger family plants such as *Hedychium* species, *Alpinia* species, and *Zingiber* species. This means rich, organic soil, copious water during hot, dry summers, partial shade and freedom from frost, although they can be grown in a heated greenhouse in hard frost areas. The flower bracts can be pale or deep green, violet or purple. The flowers, yellow to creamy pink, sometimes mauve-tipped, are clustered at the top of the 150 mm (6 in) flower stems. The whole is a bit on the blowsy side, but attractive nonetheless. The 250 mm x 150 mm (10 in x 6 in) leaves are bright green, loosely paddle-shaped and herringbone-ribbed. Propagation is by division of the rhizomes.

(Curcuma: Arabic name for yellow. *Family:* Zingiberaceae)

Curtonus *Curtonus paniculatus*, from South Africa, flowers during the summer. The 50 mm (2 in) long flowers are orange-red, tubular and star-ended, and appear on zigzag-branched flower stalks. The 1–1.2 m (36–48 in) tall and 75 mm (3 in) wide leaves are parallel ribbed like a gladiolus. Propagation is by separating the young corms in winter after the foliage has died back. Cultivation is similar to the gladiolus.

(Curtonus; 'kurtos', bent; 'onus', axis: zigzag flowers. *Family:* Iridaceae)

Cybistetes (Malagas lily; St Joseph's lily) *Cybistetes longifolia* is a South African plant of rare quality for the collector. It is like an amaryllis with its large, flower cluster of six to twenty-four, pale to deep rose-coloured, fragrant trumpets, borne at the end of the thick flower stalk. It is an autumn-flowering bulb, the flowers appearing before the 75–150 mm (3–6 in) by 12.5 mm (½ in) leaves. The leaves appear during winter. It has two tiers of leaves, one set ground-hugging, the other aerial. Although the aerial leaves die back to the bulb each year, the ground leaves extend further from the base each year. This plant likes a winter rainfall and a dry summer. It prefers a light, well-drained soil. The bulbs, which can be 225 mm (9 in) in diameter, should be planted in autumn after the plants have finished flowering, with their necks at soil level, 450 mm (18 in) apart, and left to colonise the area for some years.

(Cybistetes: 'kubisteter', Greek for tumbler; alluding to the tumbleweed nature of the plant. *Family:* Amaryllidaceae)

Cyclamen Dwarf cyclamen with their 10–15 mm (½ in) flowers borne on 150 mm (6 in) stems will, according to species, bloom at various times throughout the year. These are the hardy cyclamen, although some, particularly those from the North African countries, are frost-prone.

You are no doubt aware of the magnificent showy *Cyclamen persicum* (plate 58) hybrids, which are available in the shops and have foliage variations ranging from green with silver filigree to almost pure silver plate, with flower colours of white, picotee, pink, carmine or red. Some have a delicious boronia–lily of the valley mixture of scents. In the wild C. *persicum* is found as a dwarf, white or pink or purple, fragrant-flowered plant.

You can grow the florists' cyclamen (*C. persicum*) from seed in a heated greenhouse or heated frame in the more temperate parts of Australia and New Zealand. In the warmer parts you could grow them in the shadehouse, although temperature control over the seed is limited and therefore suspect.

Sow the cyclamen seed in seed compost such as John Innes Seed Compost, about 120 seeds to a 310 mm x 275 mm (12½ x 11 in) standard seedtray. The number of seeds can be varied according to the width of the container. The seed is covered with 4 mm (⅙ in) layer of compost. This in turn is covered by a 25 mm (1 in) layer of granulated peat, to prevent the seed from drying out. The seed is watered, and the box covered with newspaper and glass, also to prevent the seed from drying out, and kept in a temperature of 16°C (61°F) to germinate. Keep a constant check on the seed, as the glass and newspaper may draw up the young seedlings, making them weak and spindly. Remove the glass and newspaper immediately the leaf shoot emerges.

When the seeds have produced one leaf and can be handled, prick out the tubers into standard seedtrays of John Innes No 1 compost or similar, forty-five cormlets to a tray, making sure the young tuber is above ground, with its roots in the compost. Once the plant has made four leaves, plant each seedling singly into 85 mm (3½ in) pots of John Innes Compost No 2 or similar. They will flower, and the tubers can be kept for future flowers, although it usually better to sow fresh seed each year. Use only top-grade hybrid seed, available from reputable seedsmen.

Cyclamen require a free-draining soil, with organic matter such as well-rotted compost or granulated peat added to the soil, and ample water when they are growing well. They thrive in partial shade. Plant the cyclamen species as small tubers and let them naturalise.

Cyclamen are members of the primrose family, although with their swept-back blooms they look too exotic for a primrose. *C. repandum* can be found growing wild from southern France to Greece. It has fragrant, pale pink, rose pink, white-magenta or rose-magenta, spring flowers. The leaves are triangular, bordering on heart-shape. The flower stems are peculiar, as with most hardy cyclamen, since they twist spirally when producing seed heads. *C. trocopteranthum (alpinum)* from Turkey is similar to *C. repandum*, but has kidney-shaped leaves and large, candy pink, spring flowers.

The Middle Eastern *C. coum*, and its varieties, have white, pink, rose pink or purple-pink, spring flowers surmounting deep green, marbled leaves, and is an old favourite. The Lebanese *C. libanoticum* has larger 20 mm (¾ in), white-breath of pink, occasionally fragrant, spring blooms. *C. hederifolium (neapolitanum)*, from Europe, produces silver-marbled, ivy-leaf foliage and masses of pink or white, red-nosed tipped, autumn flowers, sometimes a hundred or more on old corms.

The northern and eastern Mediterranean *C. hederifolium* is unique in that it produces roots all over the corm and looks like a miniature, hairy potato. *C. cilicium*, from Turkey, follows *C. hederifolium* in flowering, but its rose, spotted-pink or white flowers have the tiniest noses, while the leaves have some of the best marbling. The European *C. purpurascens (europaeum)* is probably the most famous, because of its concentrated, penetrating fragrance. Its white or pink flowers appear in autumn. The Algerian *C. africanum*

(plate 47) is small and delicate, and bears pale pink blooms, but is not as hardy as the others. There are many miniature cultivars of *C. persicum* available, and they are grown outside in some states, having as many as fifty flowers on one tuber.

Should you buy flowering cyclamen in pots and plant them out after flowering you will find they will not do well. It could be that the area you live in may be too cold, or the tubers are tired and exhausted, and the soil you planted them in is not what they require. These tired tubers may recover the following year, given the right conditions for growth.

Plant cyclamen tubers 50–150 mm (2–6 in) apart, and cover the tubers with about 25 mm (1 in) of soil, although the planting depth of *C. hederifolium* is 50 mm (2 in); *C. coum*, 50 mm (2 in); *C. purpurascens*, 100 mm (4 in). Though the flowers of the hardy species are small, the tubers they produce eventually can be monsters. *C. hederifolium*, for example, has produced tubers up to 150 mm (6 in) in diameter.

Pot-grown tubers should be placed so that the shoulder of the tuber is just at ground level. Once the foliage has died down keep the tuber barely watered, just enough to prevent it from shrivelling and drying.

(Cyclamen: 'kyklos', circular; spiral twisting of stems. Family: Primulaceae)

Cypella The South American (Argentinian–Uruguayan) *Cypella herbertii*, cousin to the *Tigridia* species 'jockey caps', produces upside-down, three-cornered hat, centre-cupped, iris-like, 40–75 mm (1½–3 in) flowers in rust, gold and tan or burnished copper, on many-branched, 300–900 mm (12–36 in) flower stems. The three much smaller, interspaced, inner petals are rounded and pointed-tipped, like golden yellow, spotted purple, slightly inrolling, clover leaves. The 300–450 mm (12–18 in) tall, narrow foliage, is ribbed and pleated similar to gladiolus leaves. In summer cypellas form attractive flowering groups. They prefer the cooler parts of Australia, and the bulbs can be planted May–July, 50 mm (2 in) deep and 150 mm (6 in) apart, in a moist soil and open site. Water the plants well while they are growing.

The flowers and foliage will die down late autumn–winter, when the beds can be tidied up and a low-nitrogen bulb fertiliser applied as the manufacturer instructs. Propagation is by division of the corms, although cypellas are easily grown from seed, flowering usually in two years. They should be restricted to those areas where they can be controlled easily.

Varieties of *C. herbertii* exist, such as *C. h.* var. *brevicristata*, which has canary yellow flowers. Other species of cypella are: *C. herrerae*, a blue form with yellow-blue inner petals; *C. peruviana*, canary yellow flowers; *C. plumbea*, which has blue flowers feathered with shades of brown.

(Cypella: 'kypellon', a goblet; refers to flower form. Family: Iridaceae)

Cyrtanthus (Ifala lily) 'Cyrtanthus' means curved flower, and the 50–75 mm (2–3 in) tubular flowers do curve downwards. The leaves can be deciduous, evergreen or semi-evergreen depending on species.

Cyrtanthuses are available in a range of species and hybrids in flower colours such as white, pink, carmine, white-yellow, yellow-orange and orange-red. The mostly fragrant, tubular, flowers end with six small, blunt-ended petals, and appear like a fanfare of slightly down-curving

trumpets at the top, and to one side, of 200–300 mm (8–12 in), drinking-straw stems. The flowers can number from two to ten, depending on species.

C. mackenii (plate 49), with its 50 mm (2 in), fragrant ivory flowers, and its 150–300 mm (6–12 in) long, by 7 mm (¼ in) wide leaves, is the best known of the Ifala lilies, and originates from the eastern seaboard of South Africa. *C. mackenii cooperi* has late winter–spring, 37–50 mm (1½–2 in), cream-yellow flowers, and two to six leaves that tend to arch, and can be 375 mm (15 in) tall. *C. sanguineus* is a blatant autumn-red. **C. o'brienii** has bright red, scentless trumpet flowers in early spring. Its foliage is 300 mm (12 in) tall, by 3 mm (⅛ in) wide. *C. angustifolius* has red, summer–autumn, 50 mm (2 in) long, curved, tubular flowers, and narrow foliage. *C. obliquus* has 450 mm (18 in) tall, by 25–50 mm (1–2 in) wide, evergreen or near evergreen, twisted-apex foliage.

Plant the bulbs in autumn in an organic soil, or one enriched with peat. The soil should be well drained, and the site receive bright sunlight. The narrow, grass-like foliage appears first in late winter, and is followed by flowers in early summer–autumn or spring, depending on species or variety; some even flower in autumn, particularly after a good rain. This is a bonus. Plant the bulbs so that the necks are just at, or just above ground level, 75–150 mm (3–6 in) apart, depending on bulb size. You can plant them in a rock garden. Divide the clumps, when necessary, during the dormant period.

(Cyrtanthus: 'kyrtos', curved; 'andros', male flower. Family: Amaryllidaceae)

Daffodil (see *Narcissus*)

Dahlia Dahlias are natives of Mexico and Central America, and when first found by explorers were a source of great interest, as the plants provided a treasure trove of colour and design. These summer–autumn flowers can dominate our gardens, but their exotic colours do not necessarily fit into all garden schemes.

The origin of the dahlia is not known. it is thought to be a chance cross between two genera of plants resulting in a chromosome structure peculiar to dahlias; possibly *D. pinnata* was the .ginal. *D. variabilis* is a convenient name given to an original dahlia species. Ironically, the original dahlia is considered by some experts to be hybrid.

Some dahlia experts maintain that one parent of the modern dahlia is *D. imperialis (maxonii)*, a giant tree dahlia reaching to 4.5–5.4 m (15–18 ft). They also state that *D. coccinea* was the other parent, because from these two species can be produced all the known dahlia colours.

For convenience we can divide dahlias into two groups, those which produce magenta or ivory flowers, and those which produce scarlet or orange flowers. From these can be produced the startling hybrids we take for granted today.

Dahlias have been revered as garden plants for hundreds of years. The South American Aztecs were crowding their gardens with 'cocoxochtls' (water pipes — dahlias), long before Columbus. The conquistadors recorded the flowers, and in the eighteenth century Abbe Vincente Cavanillas took seeds and tubers back to Spain, where they were classed and illustrated as garden flowers.

From these humble beginnings, and cross-pollination between the species *D. pinnata*, *D. coccinea*, *D. rosea* and *D. juarezii*, today's dahlias were developed. Forms such as Giant, Medium and Miniature Decoratives; Giant, Semi, Medium, Fine-petal Miniature, Charm and Exhibition Cactuses; Nympheas; Show; Pompones (pompons); Collarettes; Orchid, Single and Double Flowered. There's a world of difference between them. **(Plates 50-62)**

Decorative Form

Formal Must be fully double for more than its radius. Petals broad, pointed or rounded.

Informal Fully double for more than its radius. Petals long, twisted, pointed or serrated and irregular in arrangement. Centre should be high and cone-shaped.

Sizes Giant: Not less than 200 mm (8 in) in diameter. Medium: Under 200 mm (8 in) in diameter, but not less than 150 mm (6 in). Miniature: Under 150 mm in diameter, but not less than 112.5 mm (4½ in). Charm: Under 112.5 mm (4½ in).

Cactus Form

Fine Petal Fully double bloom, at least radius in depth. Petals should have margins directly downwards and turning towards the midrib. Folded quills arranged regularly; no confusion. Slightly incurved or straight with high cone-shaped centre.

Semi Cactus Fully double blooms at least radius in depth. Petals broad at the base and revolute, with fold or quill two-thirds or more of their length. Regular arrangement; no confusion; slightly incurved or straight. High cone-shaped centre.

Sizes Giant: 200 mm (8 in)

Medium: 150 mm (6 in) but under 200 mm (8 in)

Miniature: 112.5 mm (4.5 in) but under 150 mm (6 in)

Charm: under 112.5 mm (6 in)

Exhibition Cactus Form

Fully double bloom, radius in depth. Long, quilled, narrow petals, arranged regularly with no confusion. Incurved preferred. False or hair petals should not show. Cone-shaped centre.

Size Not less than 150 mm (6 in) in diameter

Ball Form

Fully double blooms globularly symmetrical. Petals regular, fluted or rounded, getting progressively smaller towards the centre. Reflexing to the stem. The centre should show no depression.

Sizes Show: 100 mm (4 in) or more in diameter

Miniature: 50 mm (2 in) or more but less than 100 mm (4 in)

Pompon: 37.5 mm (1½ in) or more but under 50 mm (2 in)

Anemone Centred Form

The disc should be surrounded by eight broad, flat petals, spaced evenly. The centre should be composed of a dense collection of tubular florets, which are longer than the centre disc florets. Centre to be full and dome-shaped.

Size To be less than 125 mm (5 in) in diameter

Collarette Form

The inside, encircling row of eight florets shall be reasonably circular in outline; slightly overlapping and of good quality. Not more than half the length of the outside florets; and be of contrasting shade or hue to the florets.

Orchid Form

Similar in appearance to single dahlias, but more exotic. Petals are involute (twisted) for two-thirds or more of their length.

Peony Form

Peony-shape with two or three ruffs of petals surrounding a centre disc.

Star Dahlia Form

Two or three rows of star-shaped petals, barely overlapping or cupping a centre disc of florets.

Nymphea Form

Five or more rows of open, regular, rounded or pointed petals, depressed like a saucer with highish cone centre.

Size 112.5 mm (4.5 in), but less than 150 mm (6 in)

Fimbriated Form

Petal edges are fimbriated (clipped-looking), as one clips paper with scissors. Judged as cactus dahlias. Fimbriation must be at least three-eights away from outside.

Should you be interested in learning about dahlias, contact your local dahlia society. Your nearest Botanic Garden may have the address. You may also find that your local dahlia society has its own flower classifications, which are peculiar to that climatic area. They could have more classes then those mentioned above.

Dahlias, with their gun-barrel stems, make wonderful cut-flowers, having every shape and colour (except blue). They can be used as bedding flowers, such as the 'Coltness Gem' cultivars which, when massed, look like hundreds of spinning tops.

Dahlias can grow almost anywhere, even though they come from a hot place such as Mexico. In frost-prone areas they are planted out when all risk of frost is over, and the tubers brought inside when the stems have been cut down by the first light frost.

Dahlias like a rich, well-drained soil as their tubers rot in wet, stagnant soils. Apply well-rotted manure to the crop previous to dahlia planting, so that by the time the dahlia tubers are planted the manure will be well integrated with the soil. Do not use fresh manure as it rots the tubers.

You can grow dahlias directly from tubers, and you can take cuttings from tubers sprouted in the greenhouse in early spring. Taking cuttings from sprouted tubers allows you to increase your dahlia stock true to type. The tubers are placed in deep boxes and nearly covered with compost, and kept warm and watered. Soon green shoots will appear. These shoots, when 75 mm (3 in) tall, are cut off with a sterile, razor-type knife, making sure that a sliver of the old stem is attached to the new cutting, planted in boxes of vermiculite or peat and sand compost, and kept in good light in a 15.5–21°C (60–70°F) humid atmosphere, where they will send out roots. They are grown on, and hardened off before planting outside into their flowering beds.

Tubers are available in September, and can be planted during that month and in October, November and even into December, to provide a succession of carnival-coloured blooms during the summer and autumn. Those planted earliest will flower during the hotter part of the year.

Flood the ground well and let the water thoroughly soak in before planting the tuber. Never plant tubers into dry ground. A strong 37.5 mm x 37.5 mm x 1800 mm (1½ in x 1½ in x 6 ft) hardwood stake is driven in by each tall growing dahlia, with smaller stakes being used for smaller dahlias. Space the larger tubers 1.05 m (42 in) apart, in rows 900 mm (36 in) apart; the medium tubers 900 mm (36 in) apart, in rows

750 mm (30 in) apart; the smallest tubers 600 mm (24 in) apart, in rows 600 mm (24 in) apart. It is wise to plant them further apart in those areas known for intense humidity, where fungus spores travel rapidly from plant to plant.

Dig the hole deeply enough to take the tuber, but also to produce a small 50 mm (2 in) saucer depression to hold water. Place the tuber into the hole, with the sprouting eye close to the stake, so that the green shoots or dry stalks are at ground level. Cover the tubers with soil, and place slug bait. Plant only healthy tubers. Water the area to settle the soil around the tuber and roots. Select the strongest shoot. Once this shoot is about 200 mm (8 in) tall, nip out its centre growing point to force out sideshoots. Loosely tie the dahlia stem to the stake when the shoots are 300–375 mm (12–15 in) tall. Tie them twice or thrice to the stake during their growth, depending on their final height.

You can alternatively pinch out the growing tips once the plant has produced its third pair of leaves, to force new sturdy growth from below. The new growth should be tied to the stake when it is about 375 mm (15 in) tall. Pinch out the growing tips again to get even bushier plants, but seven or eight or even nine flowering stems should be enough. Pompones (pompons) need only be pinched out once, as with most small varieties.

If you wish to grow large blooms, then disbud. You will see more than one bud at the tip of each lateral once the plant reaches the bud stage. Disbud all but the healthiest. Also remove any laterals (young shoots) that spring from between the main stem and the leaves, thus forcing all the goodness back into the main stem. For exhibition work, when you want only a few top-class blooms, keep four or five laterals only springing from one tuber. Some gardeners do not believe in disbudding, but prefer a mass of smaller flowers.

Initially dahlias will not need much water, but they will once the shoots appear. For special cut-flowers it is best that the blooms be disbudded as you would disbud chrysanthemums, but it is not essential. Pick cut-flowers early in the morning or in late evening, and soak them up to their necks in water, as this keeps the cells turgid and flowers last longer. Flowers cut midday usually do not have the same sap turgidity. Dead or spent flowers and stalks should be removed from the plant.

Fertilise the dahlia bed using a complete fertiliser suitable for dahlias, and apply it as the manufacturer instructs. Water it in well. Mulch the root area with a carpet of well-rotted compost during the summer to conserve moisture, and feed the plants. Dahlias produce surface roots which are extremely susceptible to hoe or fork damage, so be careful of this.

Pests and Diseases

Dahlias, like most plants, are invaded by a range of pests and diseases. Pests include aphis, earwings, snails, slugs, mites, thrips and sundry leaf-eating insects; diseases include powdery mildew, stem rot, leaf spot and virus diseases. Most of the above (except virus) can be treated with insecticides or fungicides, but make sure the material you use is compatible with dahlias, as some can cause injury to the flowers. Read the label on the pack.

Mildew preys on dahlias in late summer. It covers the leaves like talcum powder. Treat with a fungicide such as Benlate (Benomyl).

Some growers in warmish areas on well-drained soils will not lift tubers each year. However, in frost-prone areas you will have to do so. Seek local expert advice. You may be told to wait until the stems have been hit by the first light frost. Cut down the stems to about 150–225 mm (6–9 in) above ground level. *Label the dahlias with their names immediately.* Then, as carefully as handling eggs, dig out the tubers, and up-end them so that surplus water drains out of the hollow stalks and allows the soil surrounding the tuber to dry out. Bring the tubers in before nightfall. Dahlia tubers are as brittle as glass, so be careful how you handle them. The tuber should always be protected from cold and frost.

You can do one of two things once the soil has dried out: You can clean the soil from the tubers, or you can leave the soil on.

Damaged tubers should be cut clean and treated with suitable fungicide to help prevent fungus disease from entering the wound. The tubers are stored in a cool, frost-free dry place, covered with sand or vermiculite. They can be stored in open, plastic bags filled with vermiculite until all respiration is finished, and then sealed. Constantly check the tubers for rot. When planting out the following season, they can be divided into two pieces with stalks and buds attached to each piece. The growing buds are formed around a collar at the top of the tuber, like the adam's apple in your throat. You must have these buds on a tuber for it to grow.

Dahlias can be grown from seed sown in seedtrays. Seed can also be sown in a garden bed, but work the soil into a fine tilth. The seed is sown spring and early summer for flowering from January on. Sow the seed in seedtrays and cover with a 13 mm (½ in) layer of fine compost. Water the seed gently. When the infant seed-leaves reach 50 mm (2 in) in height, carefully transplant them into their permanent beds. Do not waterlog them at this early stage of their lives.

(Dahlia: after Anders (Andreas) Dahl, Swedish botanist. *Family:* Asteraceae)

Dicentra (bleeding heart; Dutchman's breeches) Dicentra flowers are arranged along the stem, and hang rather like a row of miniature pink and white bats. The arching sprays reach out from the plant and the flowers tremble in the wind. They are natives of North America or east Asia.

The common *Dicentra spectabilis,* from Siberia and Japan, is 450–600 mm (18–24 in) tall, and is widely grown. Up to eighteen, 12 mm (½ in) wide flowers will appear along one stem of *D. spectabilis,* which meeans a 225–250 mm (9–10 in) line of rose pink, rosy red flowers. *D. formosa,* although short on flowers, produces the most eye-catching rose reds and pinks. *D. eximia,* from the eastern parts of North America, with its green or greyish green, 'carrot' foliage, is more tolerant of warm conditions than the others mentioned. It has 25 mm (1 in) long, rose pink, and very occasionally white, flowers.

Dicentras grow well in a moist, partly shaded area, as they object to being exposed to hot sunlight. Dicentra's tuberous roots are planted during autumn–winter for spring–summer flowering, 25–50 mm (1–2 in) deep, and 450–900 mm (18–36 in) apart. Dicentras can spread to 1.5 m (60 in) and grow as tall as 750 mm (30 in). The 'fern-carrot' foliage will die back during autumn and winter, and should be cleaned up. This tracy foliage is deliciously attractive in a border when growing well.

Dicentra is called bleeding heart because of the colour of the flowers, particularly the hanging centre-drop, which supposedly represents blood.

Do not disturb the clumps for several seasons, at least not until they have outgrown their welcome. However, it is doubtful whether you can grow dicentras successfully in northern New South Wales and Queensland. *D. spectabilis* and *D. formosa* are species worth growing. Cultivars from these have been produced: 'Bountiful'; 'Debutante'; 'Silversmith' are three examples. These and others from white, cream, golden yellow, rosy pink and dull red species are becoming more readily available.

(Dicentra: 'dis', two; 'kantron', spur; two-spurred flowers. *Family:* Fumariaceae)

Dichelostemma (ida-maia; pulchella) (see *Brodiaea*)

Dierama (fairy bells; Lady's wand) When you see these 1.5–2 m (60–80 in) stems with their graceful necks arching over a pond, bowed down by masses of rich pink or purple, hanging, tubular-bell, pointed-petalled, summer–autumn flowers, you will readily see why they are called fairy bells. They have deciduous to nearly evergreen foliage.

South African dieramas thrive in full sun or some slight shade, but will not tolerate severe frost areas. They can be grown in pots in the greenhouse in frost-suspect areas. The corms are planted in autumn 150 mm (6 in) apart, 75–100 mm (3–4 in) deep in a well-drained soil rich in humus, and copiously watered when they are growing well. Plant in clumps, and from the corm will appear narrow-sword, gladiolus-type, more or less evergreen, foliage. Appearing from the centre of the leaves come wiry stems.

D. pendulum will reach 1 m (40 in). *D. pulcherrimum* will reach 2 m (80 in). Masses of largish, 25–50 mm (1–2 in) long, 25–50 mm (1–2 in) wide, tubular-bell flowers will appear along the stem's length, causing it to bow gracefully under the weight. Root divisions may take up to two years to flower.

Dierama pendulum is the common species, producing white, pink, purple or even brick red flowers. *D. pulcherrimum* (beautiful cleft), is the giant Lady's wand, and has purple to purple-blood flowers. Hybrids are: 'Heron', deep, wine red; 'Skylark', purplish, pansy-violet; 'Kingfisher', pale pink, purple; 'Plover', pale pink; 'Port Wine', deep, wine purple-red; 'Windhover', bright rose pink.

(Dierama: funnel-shaped flower envelope. *Family:* Iridaceae)

Dietes (see *Moraea*)

Dipidax Dipidax is a name which means 'two leaves from water spring', although dipidax usually bears three leaves. It is a South African plant that has two known species, but the species most commonly grown is *D. triquetrum.* It is used extensively as a marginal plant and suits the boggy areas around the garden pool, where the bulbous gladiolus-like roots seems to adapt readily. However, *D. triquetrum* will grow in most garden soils, providing it is not in hot sunlight or exposed to wind, and watered well when necessary.

The attraction of dipidax is the six-petalled, white with blush pink, star-shaped, dark maroon-centred, yellow-throated flowers, with two crimson-purple, nectar-sighting spots at the base of the outer segments. They are borne loosely on one side of the 100 mm (4 in) lateral spikes. They appear slightly caged by the half-round, rolled,

perhaps rush-like, and semi-cylindrical, 300 mm (12 in) long, leaves, which grow above the flower spike. The leaf is narrowed to the tip, and grooved.

Dipidax dies down in summer when the bulb becomes dormant. Up to ten flowers are produced en masse, August–October, on the wiry, 250–450 mm (10–18 in) stems, each flower being 13–19 mm (½–¾ in) wide. The leaf on the flowering stem is different in that it is ground-hugging, strappish and open like daffodil foliage.

Plant the bulbs in autumn, about 100 mm (4 in) apart, 50 mm (2 in) deep in clumps. They can also be grown as pot plants.

(Dipidax: two leaves springing. *Family:* Liliaceae)

Disa (red disa) *Disa uniflora* is a 75–100 mm (3–4 in) diameter, three-petalled South African, tuberous ground-orchid of great beauty, borne on a 325–900 mm (13–36 in) stem. It is similar to our native greenhood-orchid in that the top petal is hooded, and encloses the column, anther and rostellum. However, the disa's two lower petals slope down and outwards to form a triangle with the hooded petal. The creamy yellow, pink hood is delicately criss-cross veined, and striped with crimson-pink; the lower petals are a soft, tempered scarlet to wild, deep pink. The plant has been hybridised to produce some more exciting cultivars, and these are usually easier to grow. The 200 mm x 37.5 mm (8 in x 1½ in) leaves are tufted and hug the ground.

Disa uniflora has from one to seven blooms on a stem, and is incredibly difficult to grow from seed, but should be within the capabilities of the experienced orchid-grower. It should be grown in pots, and treated as a frost-tender, greenhouse orchid, with a winter temperature of 7.2–10°C (45–50°F), even though in the ground in its native state, it can be subject to occasional frost. It expects cool nights, 7.2–10°C (40–50°F), throughout the warmer part of the year, and even on hot days in its native habitat it is bathed in a cool, early morning mist. It would appear to like a cool, buoyant, moist (not humid), airy atmosphere and seems to do well in an air-cooled greenhouse.

Grow the tubers in an acid, orchid compost containing a good percentage of granulated peat or leaf mould, but with free-draining grit added. The pots must be well crocked with broken clay pots. Disas need watering even when they appear to be dormant after flowering, as there is still activity going on. The compost must not be allowed to dry out. Thrips are a real pest of disas, but some pest-sprays can be equally damaging, so be prepared for this. Atmospheric control is most important with these orchids. There are over one hundred species of disa.

(Disa: top petal netted; Queen Disa, Swedish myth. *Family:* Orchidaceae)

Dranunculus (dragon or stink-lily) The Mediterranean *Dranunculus vulgaris* has large tubers, which produce 600–900 mm (24-36 in) tall, ten to fifteen flat-fingered, palm-shaped green leaves. The stem-base of the leaves is pale green and attractively splashed and striped with dark green markings. The leaves have short white and red-flecked veins spreading out from the midrib. From the leaf-cluster, in September–November, emerge 300–400 mm (12–16 in), wavy-edged, one-sided, pointed, funnel-shaped flowers (spathes). These spathes, which are similar in shape to the calla or arum lily, are black-purplish red, velvet-textured on the inside, and greenish on the outside. A long, blackish red, pointed, 'cat's tail' central spadix (true flower) appears out of the centre of the spathe. *D. muscivorus* has a greenish spathe.

The large tubers are planted in winter 250 mm (10 in) apart, 150–200 mm (6–8 in) deep, in a free-draining soil and partial shade, although they do grow in rocky places in their natural habitat. To attract pollinators the flowers emit a rotten meat smell, which attracts blowflies that carry pollen from one bloom to another. Should you want to grow these bulbs, plant them well away from the kitchen window.

(Dranunculus: name used for a curved rhizome-rooted plant. *Family:* Araceae)

Eminium The Turkish–Syrian *Eminium inortum* is an early-flowering arum, with an open funnel, top-pointed, hooded-tip, plum to black-purple 'flower' spathe, with central, 'black-pencil' spadix (true flowers). The spathe emerges from large, wavy-edged, pale green ground leaves. The spring-flowering Central Asian *E. albertii* has a 150–200 mm (6–8 in) long, pale green outer, or variegated purplish and black-purple-red, spathe and spadix. The spathe tube is swollen below. Plant the tubers 150 mm (6 in) deep and 225 mm (9 in) apart in a free-draining soil, and keep dry during summer. Eminiums could lend themselves to pot culture using sandy, open compost.

(Eminium: name given by Discoroides. *Family:* Araceae)

Endymion (English–Spanish bluebell) Many gardeners naturalise bluebells in borders, and when they flower in spring, pick bunches, which they arrange in pots, filling the house for days with sweet-sharp bluebell fragrance and blue-mauve reflection.

The large, 50 mm (2 in) bulbs are planted in clumps in shrub borders or deciduous woodland, where they receive shade and adequate moisture, supplied either naturally or by the gardener. However, if they find conditions much to their liking, they can quickly turn into a garden pest, particularly in rock gardens. Therefore, plant them where they can be contained.

Endymion non-scriptus, the English bluebell (syn. *Hyacinthoides non-scripta*; *Scilla non-scripta*), likes cool, moist conditions, and a neutral to acid, rich organic soil. There are attractive varieties and cultivars: *E. n-s. alba*, white bells; *E. n-s. alba major*, large white bells; *E. n-s.* 'Blush Queen', light pink; *E. n-s. rubra*, reddish blue. The fresh green, glossy foliage is strap-like and nearly as tall as the flower raceme.

Endymion hispanicus (syn. *Hyacinthoides hispanica*; *Scilla hispanica*) **(plate 66-7)**, the Spanish bluebell, seems to be more adaptable to the Australian climate. *E. h.* 'La Grandesse' is a large white form; *E. h.* 'Blue Queen' is a delicate light blue; *E. h.* 'Sky Blue' speaks for itself; *E. h.* 'Queen of the Pinks' has large, deep pink bells; *E. h.* 'Arnold Prinsen' is rose pink; *E. h.* 'Salmon Queen' is salmon pink. The deep green, glossy foliage, which continues growing after the flowers have finished, is strap-like, and nearly as tall as the flower raceme.

Plant the bulbs February–March 50–100 mm (2–4 in) deep, 100–150 mm (4–6 in) apart, depending on bulb size, in groups of about twelve. They will rapidly colonise an area. The English bluebell will send up 250–350 mm (10–14 in) stems, on which are borne, deliciously fragrant,

12 mm (½ in) bell-shaped flowers. They are loosely carried mostly on one side of the curved-topped, flower raceme. The Spanish bluebell, which can grow to 500 mm (20 in), has up to fifteen, larger, 17 mm (¾ in), bells, loosely borne on both sides of the flower raceme.

Top-dress the bulbs, if planted in open soil, with well-rotted leaf mould or well-rotted horse or cow farmyard manure. A dressing of a low-nitrogen bulb fertiliser can be applied to woodland bluebells.

(Endymion: mythological woodland character. *Family:* Liliaceae)

Eranthis (winter aconite) The flowers of *Eranthis hyemalis*, winter aconite, emerge from the ground like bright yellow mushrooms. They develop into large, star-shaped buttercups and hug the ground with their segmented, green-ruff leaves, rising only 80–150 mm, (3–6 in) above soil level.

Winter aconites grow happily in cold climates, and with snowdrops, chionodoxas and early tulips make a heart-warming sight in late winter.

Winter aconites can be planted to naturalise in deciduous woodlands. They can also be planted in borders or in a rock garden, but have to be contained as they may get out of hand.

Mass plant the tuberous roots about 50–80 mm (2–3½ in) deep, 50–100 mm (2–4 in) apart, depending on bulb size, in February–March. They can be grown as pot plants, and form pleasing, low-effect groups. The best known winter aconite is *E. hyemalis*. *E.* x *tubergenii (tubergeniana)*, (E. *hyemalis* x *E. cilicicus*), is a hybrid of good vigour with golden, larger blooms. The 'selection' cultivar *E. t.* 'Guinea Gold', with deep, buttercup yellow, green-bronze foliage, is popular. *E. cilicicus (cilicica)*, a near Eastern species, has larger, more open, 'buttercup' flowers and smaller ruffs.

(Eranthis: 'er', spring; 'anthos', flower. *Family:* Ranunculaceae)

Eremurus (giant asphodel; foxtail lily) Eremuruses are natives of Turkey, the desert areas of Lebanon, Jordan and through to western China. They thrive in a well-drained, sandy soil, rich in humus, with a covering of sharp sand during the dormant season to keep them reasonably dry.

The roots can be planted 100–150 mm (4–6 in) below ground, 450 mm (18 in) apart, and allowed to spread into larger clumps before dividing them. The roots are a mass of white, fleshy stars, which seem to stay close to the surface, not going beyond 150 mm (6 in) deep into the ground.

Eremurus leaves are narrow, triangular in section, tufted, strap-like, and similar to red-hot poker foliage. But the flower spikes are spectacular, as they lift off like yellow-pink rockets from the dry-looking foliage. They need ample water during the growing season, and should be planted in a frost-free area. They also need protection from slugs, which attack the young shoots. The translucent, pink-flowered forms can be squat bell-shaped, six-petalled, star-ended, with yellow star centres, and a reddish stripe going down the centre of each petal.

E. stenophyllus (E. bungei) is a yellow form that thrives in mountainous areas. It likes a well-drained soil, with full sun and protection from wind. It is a small species reaching a maximum height of 700 mm–1.5 m (28 in–5 ft), and bears a raceme of intense yellow flowers. *E. olgae* is white-

flowered with a pink stripe on each petal. *E. lactiflorus* is white-flowered. *E. cristatus* is reddish with a white outer stripe. *E. robustus* is a giant, and can climb from 1.8 m (6 ft) to 2.4 m (8 ft). It bears masses of 'corncob', delicate pink blooms, arising from blue-green leaves.

There are many hybrids available: 'Shelford', orange-buff; 'Warei', breath of orange; 'Moonlight', pale yellow; 'Rosalind' bright pink; 'White Beauty', white; 'Isobel', pink-orange; and others. Eremuruses flower mostly from spring on, and a constellation of blossoms will appear up each stem.

However, once flowered, the dying eremurus flower spikes and foliage look untidy, so be prepared to accept the gorgeous flowers for what they are worth. In herbaceous or mixed borders, gardeners should attempt to plant tall summer flowers, which will hide the scruffy foliage.

(Eremurus: 'eremos', solitary; 'oura', tail. *Family:* Liliaceae)

Erythronium (dog's tooth violet) With their spring-flowering, star-shaped, swept-back, reflexed turk's cap blossoms, dog's tooth violets remind one more of hardy cyclamen than of violets. They prefer a cool, moist climate and a well-drained soil.

The species *E. dens-canis* ('dens-canis' meaning dog's tooth: the shape of the corm) has been cultivated for centuries, and its rose pink, purplish flowers and marbled mottled foliage are great favourites. The leaf mottling of some forms alone is a feature worth cultivating. They can be naturalised in rock gardens, where, clustered, they seem to nod. The 'canine-tooth' corms are planted in autumn, 50–80 mm (2–4 in) deep and 50–75 mm (2–3 in) apart. *E. dens-canis* can be had in shades of pink, red, reddish violet and white.

There are many other species of dog's tooth violet, which are mainly of North American origin. Cultivars of these, and the European dog's tooth violet, have been bred. *E. californicum* (Californian fawn lily) has brown mottled leaves and creamy white flowers, or yellowish with orange-yellow base. *E. revolutum* (Californian trout lily) is highly prized for its mottled, marbled, blotched foliage. The gorgeous cream-pink or dark pink blooms are borne on 250–450 mm (10–18 in) stems. The hybrids are more prized than the species, and range in colour shades from yellow with gold centre ring, to yellow with brown centre ring. *E. r.* 'White Beauty', is arctic white with a pale yellow to chocolate zone.

E. tuolumnense, with its plainer leaves and golden yellow, green-based, six-petalled, star blooms, borne on 250 mm (10 in) stems (with some deeper green on the outside petals), is becoming a great an attraction as any dog's tooth violet. *E. montanum* is a white form. *E.* 'Jeanette Brickhill' is a beautiful, soft pink with red stigma-style anthers. *E. americanum* has powdery, mottled leaves and mimosa yellow flowers borne close to the ground. *E. grandiflorum* has creamy white, golden yellow, green-streaked outer, star-shaped petals.

Dog's tooth violets are best left undisturbed for some years, after which they can be separated by division of the corms. The corms are planted in autumn. The American forms can be planted 75 mm (3 in) deep, 75–150 mm (3–6 in) apart. Some increase readily from seed, and many of the American dog's tooth violet species are propagated in this fashion. In three, four or five years they will produce flowers.

(Erythronium: 'erythros', red; European flower colour. *Family*: Liliaceae)

Eucharis (Amazon lily; should be called Colombian Andes lily) *Eucharis grandiflora* will tolerate only the humid subtropical parts of New South Wales and Queensland. However, it is grown by greenhouse owners in a winter temperature of 15.5–21°C (60–70°F), and summer temperature of 26°C (80°F); the greenhouse being shaded if necessary.

Its 100 mm (4 in) wide, waxy, hospital-white, green-centred, fragrant, narcissus-like flowers, are a joy to behold and inhale. The slightly pendulous flowers erupt in small clusters of up to four, or more, at the top of straight, 600 mm (24 in), daffodil-like stems (scapes), and reach high above the wide, shiny evergreen, oval to paddle-shaped foliage.

Plant the bulbs 50–75 mm (2–3 in) below soil level in a well-drained soil, 250–300 mm (10–12 in) apart, shaded, out of the hot sun, but in an area which receives enough ambient light for the flowers to thrive. They need ample water during the growing season, but not so much during their dormant season. However, being evergreen they must never be allowed to wither. When, during the slow-growth period the leaves have reached full size but no new leaf growth is showing, water can be withheld until the leaves show the very first signs of wilting. They must never be allowed to wither. Then the plant should be flooded with water. You can do this a number of times during the seasonal four to six weeks of slow growth. Afterwards normal watering and feeding can be resumed. Propagate by careful removal of bulb offsets. Eucharises do not like having their roots disturbed once they have become established, and can react by not flowering.

Eucharises prefer a well-drained fibrous loam soil, which has well-rotted farmyard manure, peat, plus sharp sand added. When potting Amazon lilies crock the pots for drainage and use a compost of two parts fibrous loam, two parts gritty, sharp sand, one part well-rotted leaf mould, one part granulated peat, plus charcoal. Eucharises can be grown in the shadehouse, one bulb to a 250 mm (5 in) pot or more in 250 mm (10 in) pots.

Amazon lilies will flower in spring and occasionally during the summer. Feed the plant with liquid cow manure, diluted to the colour of weak tea, when the flower scapes (stems) appear. It is practical to restrict casual flowering to allow the plant a period of rest, and thus it will produce good flowers the following spring.

E. grandiflora has slightly drooping, 100–125 mm (4–5 in) flowers with as many as five blooms on one 500–600 mm (20–24 in) leafless stem. *E. candida* is a smaller, white-flowered version, but still popular.
(Eucharis: Greek, meaning graceful attraction. *Family*: Amaryllidaceae)

Eucomis (pineapple flower; pineapple lily) The reason the eucomis is called pineapple lily is because the scape of flowers, with the 'pineapple' (coma) leaves above, resemble a pineapple. The species widely grown are *E. cosmosa (punctata)* **(plate 64)**, *E. pole-evansi* **(plate 65)** and *E. bicolor*. Eucomises originate from South Africa, where more than ten species can be found proliferating the hills and meadows.

Eucomis cosmosa has 12–25 mm (½–1 in) long blooms, borne in a 300 mm (12 in) long cluster

with characteristic spotting on its flowers and leaves, which distinguishes it from *E. bicolor*. *E. bicolor* has distinct purple margins to its petals; its flowers borne in a 75–100 mm (3–4 in) cluster. *E. pole-evansi* is taller, 900 mm–1.5 m (36–60 in), with a 600 mm (24 in) flower raceme.

The 75–100 mm (3–4 in) wide bulbs are planted with their noses just at, or 50 mm (2 in) below, soil level and 225 mm (9 in) apart. They will produce sturdy, cylindrical, 300 mm (12 in) leafless stems (scapes) from which will burst scores of star-shaped, white-green flowers, with their characteristic purple anthers. *E. pole-evansi* has much taller flower stems. The dead flowers remain on the stalk and still look attractive. The spring green, wavy-edged, 550 mm x 90 mm (22 in x 3½ in) strap-like foliage, with prominent midribs, forms proud rosette-tufts around the flowering stem.

Plant the bulbs in autumn for summer flowering (some recommend September), in an open site, in a rich, organic, free-draining soil. Provide ample watering during the growing period. They will need a bulb fertiliser applied in November. Stock can be increased by dividing the bulbs after the foliage has died back. Seeds will take a long time to flower.

Pineapple lilies can be used as cut flowers, but there seems to be some objection to the odour; the plant has a peculiar smell. Eucomises, being spiky, make fine pot plants, planted in John Innes Compost No 1, or similar. They can be grown massed in garden beds, where they resemble green hyacinths.
(Eucomis: 'eukomes', beautiful head of leaves. *Family*: Liliaceae)

Ferarria (syn. *Facaria fascicularis*; *Ranunculus kochii*) *Ferarria crispa (undulata)*, the 300 mm (12 in) tall, spring-flowering black iris, produces crinkled-edge, greenish black, triangular, iris-like, 50 mm (2 in) wide flowers, which have an unpleasant smell. Plant these South African corms 100–200 mm (4–8 in) deep, 150 mm (6 in) apart, March–May. A collector's plant.
(Ferraria: Giovanni Ferrari seventeenth century botanist. *Family*: Iridaceae)

Freesia Freesias **(plates 68-9)** are grown extensively as cut-flowers, because of their sweet, fragrant and exquisite pastel orange, red, yellow, blue and white, split-edged, 40–50 mm (1½–2 in), fairy-trumpet flowers.

These South African natives have adapted to growing in Australia. One mostly sees them growing in pots, but they also make splendid specimens in a well-drained border, where late winter–spring, their wiry, 150–600 mm (6–24 in) stems will produce a mass of flower trumpets on one side of the stem.

Plant the corms in autumn (even in June) 50–75 mm (2–3 in) deep and 50 mm (2–3 in) apart in a sheltered, sunny part of the garden, a site which receives ample moisture. Leave the plants to colonise, which can take up to four or five years, or flower production deteriorates. The clump can be lifted, and the smaller, healthier corms replanted. Destroy unwanted corms.

Although the species *F. refracta (alba)*, with its white inside trumpet with golden buttercup blotch, and externally a breath of violet, is widely grown, there are many colourful cultivars (hybrids) available in a wide range of pastel shades. Some deeply saturated coloured cultivars are listed in the bulb catalogues.

If you wish to grow freesias from seed use the following method. Sow the seed from February–March onwards. Blue freesias seem to be most popular, followed by yellow, red and white. Stagger the sowings; blue first, yellow seven days after blue, red fourteen days after blue, and white twenty-one days after blue. Theoretically they should all flower together.

Some gardeners practice chitting the seed before sowing. This is done by suspending the seed in a muslin bag in tepid water for twenty minutes each day until the seed germinates. The seed is then mixed with moist granulated peat and sown in pots of John Innes Seed Compost, about six seeds to a 125 mm (5 in) pot, or one seed to 1280 mm² (2 square inches).
(Freesia: in honour of Freidrich H. T. Freese. *Family*: Iridaceae)

Fritillaria Fritillaries extend from France, through the Balkans to the steppes of the USSR, and some species move on into Siberia. Of the hundred or more species, the most famous are *F. nigra*, the so-called black fritillary; *F. meleagris*, snake's head fritillary; and *F. imperialis*, the stately crown imperial.

It is difficult to get most spring-flowering fritillaries to adapt to Australian climatic conditions. They are cold plants and need a cold winter and a cold spring to establish their flowering period. We are, therefore, discussing the growing conditions one would expect to find in Tasmania, the high areas of New South Wales, South Australia, Victoria and southern parts of New Zealand. They thrive in semi-shaded sites, in a deep soil, rich in peat or compost. However, certain Middle Eastern species, such as *F. persica* and its hybrids, grow naturally in Meditteranean-like conditions.

F. nigra is not a black fritillary as the name would suggest, but is a brassy bronze red. The 25 mm (1 in) flowers are nodding, bell-shaped, angular egg-cups, blotched brown, with six-pointed, serrated petals, enclosing a yellow centre.

The snake's head fritillary, *F. meleagris* grows in deciduous woodland and reminds one of a small snake poised to strike. The nodding, egg-cup size, angular bell-shaped, six-pointed-petalled flowers are 25 mm (1 in) long. They hang singly from slender stems, the stems reaching to 250–325 mm (10–13 in) in height, and are suited to rock gardens. They come veined, chequered and blotched in red, reddish purple, white, white-green and violet. Plant bulbs in March 90 mm (3½ in) deep, 100–125 mm (4–5 in) apart in a well-drained soil, rich in humus. They can be grown in pots in the shadehouse.

F. imperialis, 'Tears of Mary' is named because the tips of the petals seemingly large, nectar teardrops from watching the crucifixion of Christ. 'Crown Imperial' is so called because of the crown (whorl) of blooms that surrounds the top of the high stem, looking like a floral crown. To see these stately blooms borne on solid stems 600–900 mm (24–36 in) above the leaves, but below the whorl, reminds one of an elegant, multi-lanterned, miniature street lamp. The mid-late spring blooms are hanging clusters of angular bells; pointed and blunt-toothed. They are available in orange, yellow, lemon yellow, and orange-red. The bulbs may have to be lifted from the ground in certain areas as they tend to rot from summer rain. They tolerate a moderately limy soil. Plant the up to 150 mm (6 in) diameter bulbs 75–100 mm

(3–4 in) deep, 150–300 mm (6–12 in) apart, in autumn.

Some people are offended by the garlic odour of the Crown Imperial, but the art is to plant them where they can be seen, but not smelt.

F. persica, particularly the cultivar *F. p.* 'Adiyaman', bears up to twenty-five, 20 mm (1 in) reddish brown hanging, pointed-petalled bells on 1 m (40 in) stems. *F. pyrenaica*, 300–600 mm (12–24 in) tall, has one to three bell-shaped flowers, recurved at the tips, with polished purple-red outside and copper-yellow inside.
(Fritillaria: 'fritillus', a dice box; flower veining. *Family:* Liliaceae)

Funkia (hosta; plantain lily: see *Hosta*)

Gagea A European–Middle Eastern, 50–250 mm (2–10 in) tall, early flowering, lily plant, with one or two narrow, basal leaves. The rest are 2 mm (¹/₁₀ in) wide, half-tubular, and arranged almost opposite each other on the stem. The mimosa yellow, pale green-tipped, pointed, narrow, six star-like petals of some species of gagea are startlingly lovely. *G. lutea* has one, rarely two basal leaves, and bears five to seven, 18–25 mm wide, bright yellow blooms with six spreading petals on 100–250 mm (4–10 in) stems. There is a broad band of green down the outside of each petal. The yellow, star-like appearance of *G. fibrosa* is impressive. *G. bohemica* has fine rush-like leaves, and blunt, open petals. *G. fistulosa* has slightly closed, mimosa yellow petals. Grow them in pots using natural division of the bulbs in autumn, in well-drained sandy compost, to contain them, and destroy any surplus bulbs.
(Gagea: Sir Thomas Gage, eighteenth–nineteenth century botanist. *Family:* Liliaceae)

Galanthus (snowdrops; milk flower) Snowdrops will appear in winter, out of nowhere, pearling the earth with their small, 12–20 mm (½–¾ in), delicate, white, pendulous, flowers. There is no flower more welcome than these European and western Asian handmaidens of winter. In Australia snowflakes are more commonly grown, but they are larger than the true snowdrop. Snowdrops have adapted to the winter conditions found in Tasmania, the hills of Victoria and New South Wales, and in New Zealand. They appear winter–spring.

Snowdrops make uncomplaining rock garden plants, where they should stay for a long time to colonise. The flowers, which hang like six-petalled fairy bells from 100–200 mm (4–8 in) stems, are snow white, some with green or yellow-tipped petals.

Snowdrops are best divided during flowering or immediately after; by doing so you will find they will recover quickly. The blue-green, narrow, strap-like foliage will continue to produce food for the bulbs. Treat them gently. Plant the bulbs when they become available in autumn, in moist, not wet, soils, 50 mm (2 in) deep, and 75 mm (3 in) apart. Do not use fresh farmyard manure near snowdrop bulbs as it rots the bulbs.

The common snowdrop, *G. nivalis*, has three outer, spreading, snow white petals, with three green and white, inner petals that form a tube. There are many superb cultivars of *G. nivalis*, many with larger flowers: *G. n.* 'S. Arnott'; *G. n.* 'Magnet'; and *G. n.* 'Atkinsii'. *G. elwesii* has 18.75 mm (¾ in), white, waxy flowers, with meadow green inner petals, orange-yellow stamens and twisted leaves. *G. n. rachelae* is a

superb autumn-flowering form. The last time I saw it was a lifetime ago in the alpine house of a famous garden, where I worked as a young man. *G. n. reginae-olgae*, 180 mm (7 in), is another autumn form. There is a double-petalled snowdrop *G. n. flora-plena (pleniflorus)*, which is attractive, but not as delicate as the others. Snowdrops make exquisite cut-flowers.
(Galanthus: 'gala', milk; 'anthos', flower. *Family:* Amaryllidaceae)

Galtonia (*Hyacinthus candicans*) (spire lily; summer hyacinth) The South African *Galtonia candicans* is called spire lily because it produces flowering stems that tower like a church spire above the almost lax, grey-green, 600–900 mm (24–36 in) tall, 25–50 mm (1–2 in) wide, pointed, strap-like leaves. In summer, on each 900 mm–1.25 m (36–50 in) flower spike, are borne upwards of fifteen loosely spaced, drooping, segmented, trumpet to bell-shaped, sweetly fragrant, pure white, tipped-green blooms. These flowers grow 37.5–50 mm (1½–2 in) long, and 10 mm (½ in) wide at the mouth. *G. princeps* has white bells with a breath of green.

Galtonias like a rich soil and a sunny, well-drained position in the garden. Plant them autumn–winter, when they are dormant, 125 mm (5 in) deep and 200–250 mm (8–10 in) apart, in clumps of ten, and leave to flower undisturbed for years. Some people do treat them, depending on climatic conditions, like gladioluses, and lift and store each year. Slugs love them, so be careful.

You can naturalise galtonias in turf areas, but the floppy foliage and tall flower spikes make them oddly at home. Perhaps they are best planted in a border.

Propagate by division of the bulbs, or by sowing seeds which produce bulbs reasonably quickly. Cut leaves, still on the plant, have been known to produce bulblets at the cut-scar tissue.
(Galtonia: F. Galton, of fingerprint identification fame. *Family:* Liliaceae)

Geissorhiza (wine cups; sequins) The South African, 150–300 mm (6–12 in) flower-stemmed *Geissorhiza rochensis*, with its 120 mm x 10 mm (4¾ in x ½ in) wide, grass-like foliage, tapering to a point is aptly named as wine cups. The 30 mm (1½ in) wide, six-lobed, velvet-textured, deep violet petals form a wine-glass, which encloses a claret red wine stain that has a thin, white froth-line at the top. However, the wine has been carelessly poured, as one can see flecks of white sediment on the sides.

The pea-sized corms will increase rapidly and produce five or more flowers on one flower spike. Plant the corms January–May 25 mm (1 in) deep, 100 mm (4 in) apart, in a well-drained soil, but contained, as they could spread and become a weed menace. Geissorhizas are frost-tender and should be grown in a sheltered, sunny spot. They can also be grown in pots, in John Innes No 1 Compost or similar; four or five corms to a 125 mm (5 in) pot, in a spot where shelter can be provided easily. *G. splendidissima* is a vivid, sea-blue. *G. furva*, is fresh-blood coloured.
(Geissorhiza: 'geisson', fibre tiles; 'rhiza', on root. *Family:* Iridaceae)

Gelasine The summer-flowering South American *Gelasine azurea* is a rare member of the iris family, growing to a height of 300–900 mm (12–36 in). It is similar in appearance to some of the larger *Romulea* species. The leaves are

parallel-veined and pleated like a bearded iris. The long flowering stem is topped by many branches, which arise in the axils of the stem and stem leaves. Many 37–50 mm (1½–2 in) wide, six-petalled, dark blue, whitish base, speckled, blue-black star flowers erupt from these branches. The flowers last only a few hours, but many flowers are produced over a month's flowering period. It is a vigorous seed and corm producer, and should be contained in post.
(Gelasine: 'gelasinos', smiling dimples; flowers. *Family:* Iridaceae)

Geranium Geraniums are not considered as bulbs, but there are a few species that produce tuberous roots. One is the Mediterranean, spring-flowering *G. tuberosum*, with its 225 mm (9 in) tall, carrot-leaved foliage, and mauve-pink 12.5 mm (½ mm) open flowers. *G. t. charlesii* is similar. The Asian *G. transversale* has lilac-pink flowers. Grow them in 125 mm (5 in) pots, in John Innes Potting Compost. Transfer to 250 mm (10 in) pots when necessary. A collector's plant.
(Geranium: 'geranos', crane's bill; flower shape. *Family:* Geraniaceae)

Gladiolus (sword lily) Gladiolus, which as a genus is far-ranging, from South Africa to Europe and the Middle East, can be divided into summer rainfall or winter rainfall clans. We all know the large-flowered florists' gladioluses, a mass of open, flared, double or sometimes hooded blooms that are available around Christmas. These hybrids are mostly bred from South African Eastern Cape species.

Gladioluses (**plates 71-77**) are easily grown from corms and there are hundreds of cultivars available in all colours of the rainbow. As a cut-flower they have no peer and are best used when the bottom florets are flaunting, but the top florets are closed. In this way they flower for a longer time in the house. The corms are also classified as early or mid or late flowering, and this is something you have to find out. Gladioluses can also be 'secund' — flowers forming and facing one way on one side of the spike — or 'distichous' — flowers spaced alternately and facing opposite ways on the spike.

Gladiolus societies, formed to advance the knowledge and culture of *Gladiolus* species, abound in Australia and New Zealand, and your local Botanic Garden or Parks Department may know the address of some close to your home.

Gladioluses can be divided into four convenient groups: dwarf (or *nanus*); *primulinus*; butterfly, and large-flowered.

Dwarf gladioluses are short-stemmed, growing no more than 600 mm (24 in). They are available in every shade and hue, and are useful as pot plants.

The *primulinus* hybrids are cobra or hooded gladioluses. The top petals curve forward to hood the centre stamens, similar to the top petal of a snapdragon (antirrhinum). The flowers, being 50–75 mm (2–3 in) across, are arranged irregularly up the stem, but are still beautiful as cut-flowers. The shades and hues of *primulinus* gladioluses are Renoir pastels in variety.

The petals of the butterfly gladioluses are ruffled and have trailing sections like a butterfly settling on a flower. They also have the added attraction of different colour blotches on the throat of the flower, such as fresh blood against butter yellow.

The 200–250 mm (4–5 in) wide, large-flowered

gladioluses, thirteen or more blooms alternately facing either side of the 1.2 m (48 in) flower stems, are those we see for sale at the florist's from late spring on. The colours are solid or shades and hues of white, yellow, red, pink, blue, lavender and purple-black. They are readily available and easy to grow.

Gladioluses are planted in an open, sunny position protected from hot, drying winds, which could scorch the young, growing tips. They grow best in a free-draining, sandy loam. However, should you not have such a soil I suggest you scoop out a hole and lay the corms on sand. The soil must be free-draining, as any stagnant water will rot the corms.

Gladioluses do well in what we call a second-season soil. I have worked in gardens where gladioluses were grown as cut-flowers. The soil in the previous year had been manured and a crop of potatoes grown in it. In due season gladioluses followed the potatoes. Potatoes do wonders in breaking up the soil, and the gladioluses did well. Gladioluses do not like fresh manure or acid soil.

Plant healthy, plump, high-crowned gladiolus corms 125–150 mm (5–6 in) deep, in a sandy loam soil; and 75–100 mm (3–4 in) deep in a loam soil (the large corms being planted deepest); and 100–375 mm (4–15 in) apart, depending on size of corm and for what purpose the plants are to be used, e.g. cut-flower, garden flower or showing. Generally, they are planted 150 mm (6 in) apart. 'Show' gladioluses are planted wider apart to get larger blooms; cut-flowers are planted closer together; and flowers grown for your personal garden show are arranged to look attractive. The rows are 600 mm (24 in) apart. You could have the rows closer, but remember you have to work in between them.

You can add a complete fertiliser at the manufacturer's recommended rates for gladioluses, which can be scratched into the top surface but kept from contact with the corms. Gladioluses will need staking, which is best done when planting the corms, to avoid spearing the roots once the corm has established.

Start planting in May, then at two-weekly intervals to early September. However, Australia has such a variable climate that the time to plant, and indeed the depth to plant, will depend on local conditions. As a general guide, in coolish areas they can be planted in August, but only if the risk from frost damage is over; in warm areas May to September. Plant deepest in hot climates, and shallower in cool climates. In the north gladioluses can be decimated by seasonal storms, and planting should be so arranged to avoid, as far as is possible, the likelihood of gladioluses being in flower when these storms occur.

Spread out your gladiolus corms in a shed so that you can watch them. If you see green life peeping from the top of the corms, plant them. Do not delay as the corms need soil in which to grow. Soon a sword point will stab through the soil. It is the gladiolus leaf, and it should be carefully avoided when weeding around it. Damage to this shoot could have a disastrous effect on the flowers. Gladioluses take up to three months to flower.

The corm will need water as it grows. You must avoid splashing the flower petals once they are in bud and flower. Cut the flowers for the house when the bottom flowers have turned into small dishes, but the top flowers are still tight. The plant will need to be tied securely, to the stake, but not throttled.

Leave the foliage to die back naturally after the

gladioluses have finished flowering. You must cut off the dead and dying flower stems to force all the goodness from the leaves back into corms. Cut off and burn the dried foliage once it has turned yellow, but leave 75–100 mm (3–4 in) of leaf stem still attached to the corms. Dig up the corms, taking great care not to spread unwisely the spawn attached to the corm. Store the corms in a dry, airy place, and soon you can clean off all the extraneous matter. Once dry and clean the new corms and cormlets can be stored for future use.

In Summary
1 Buy good quality corms, free from disease or insect infestation.
2 Check the outer husk of the corm before planting, to ensure that no rot has taken place during the dormant period.
3 Place the corms into a special gladiolus insecticidal dip before planting, to kill thrips and other insects. *Take care: read the label on the packet and follow the instructions.*
4 Plant the corms in a well-drained soil in a sunny position.
5 Water the corms only when necessary during their early, winter growth, and then only during dry spells.
6 Thrips will attack at the three-leaf stage. Spray with a suitable systemic insecticide or complete gladiolus spray at the recommended intervals.
7 Give the plant copious water, if no rain has fallen, when it has reached the four-leaf stage. A complete gladiolus fertiliser at this stage will be beneficial, but keep it away from the corms. Tie the plant securely to the stake but do not over-restrict.
8 Dig up the plants six to eight weeks after flowering has finished. Cut off the dead foliage and burn it. Dry out the corms for a few days. Dip or dust them in a suitable insecticide and fungicide for the recommended time. Remove any spoil, and upturn plant on a sieve or similar.
9 Clean off old corms and cormlets about a month after drying, and store the corms and cormlets in non-sweating bags in a cool, dark, airy place for the following year.
10 Regularly inspect the corms and discard any that are suspect.

Gladiolus species
Many hybrids have been derived from G. cardinalis x tristis species, (now known as G. x colvillei — deep pink and yellow blooms). G. cardinalis, a native of South Africa, has blatant red, white-tongued petals, with a large red petal always at the top. G. cardinalis has been used extensively in cross-breeding. G. tristis is a pale to lemon yellow, hardy species, sometimes tinged red, with a delicious evening fragrance. G. byzantinus is red-purple with some white blotching and white streaks on the lower petal. G. alatus is the miniature, rock garden gladiolus, is red with a yellow base. G. natalensis (psittacinus), the parrot gladiolus, flowers in dense groups of red-yellow blooms; can become a pest. G. primulinus is buff, mimosa yellow, thought to be a form of G. natalensis. G. papilio, white, flushed pink with yellow inside, is a parent of the butterfly hybrids. G. orchidflorus has green-yellow, purple, red-brown, scented, orchid-like blooms. G. italicus (segetum), a Mediterranean species, has reddish lilac-blue flowers. From two hundred or so species world-wide, twenty-five thousand cultivars (garden varieties) have been bred.
(Gladiolus: from 'gladius', sword-leaves. *Family:* Iridaceae)

Gloriosa (glory lily; flame lily; climbing lily)
Natives of African and tropical Asia, these glorious orange-scarlet, or sometimes pure yellow, herbaceous, climbing lilies will clamber over 1–2.4 m (3–8 ft) walls and trellis. They clamber by using curly tendrils located at the tips of their tapered leaves.

G. rothschildiana (plate 78) is a popular gloriosa. The six, 75 mm (3 in) long, narrow, wavy-edged, red and yellow petals are swept back from the centre ovary, as if being forced by a strong wind. Even the centre style, which usually sticks straight out in most plants, is bent almost at right angles. G. r. var. citrina is a yellow, flushed hybrid.

G. superba, which is more tolerant of cold weather, has yellow, orange-red 'turk's cap' flowers, with 10 mm (1/2 in) wide, non-wavy flower margins. G. simplex (G. virescens) has startlingly red or green-yellow flowers, and 20 mm (3/4 in) wide, wavy-margined petals. There are many other species both dwarf and tall, some with clinging leaf tendrils, others without. Recent work carried out on the classification of gloriosas suggests that many of the varieties of gloriosa are forms of G. superba.

The thin sausage, or finger-like, tubers should be reasonably long, 125–150 mm (5–6 in), to ensure flowering. They should be planted horizontally, June–September, with the top of the tuber 30 mm (1-1/4 in) below the soil surface, well apart, 900 mm – 1.2 m (36–48 in), to allow for top foliage spread, in a warm, sheltered spot, or shadehouse (or under glass in the cool areas), for flowering from late summer onwards. The tubers are reputed to be poisonous so keep them away from children and pets.

Gloriosas are greedy plants and like a rich, well-drained soil containing plenty of well-rotted humus. They like ample water when growing. Keep up the watering, provided no apreciable rain has fallen, as long as the foliage remains green. The tubers should be planted when all fear of frost is over. You can apply a liquid fertiliser when the plant is growing well, just before flowering.

Gloriosas make ideal pot plants providing you use a large pot, and place one 125 mm (5 in) tuber to a 150 mm (6 in) pot, or one 150–175 mm (6–7 in) tuber to a 225 mm (9 in) pot. You must also provide suitable climbing features for them to clamber on.

You can lift the tubers once the foliage has died back. You can divide the tubers in autumn, but be careful as the tubers are brittle and damage easily. Separating tubers is an easy process when repotting, as the tubers divide naturally while growing. When dividing the tubers cut exactly through a tuber-joint (knuckle) making sure that a portion of the joint remains at each end of the tuber. This ensures that shoots will rise from each end of the tuber. Store tubers in dry peat moss or dry sand.
(Gloriosa: full of glory. *Family:* Liliaceae)

Glossodia (waxlip orchids: see *Orchids*)

Gloxinia (Sinningia speciosa hybrids) These 75 mm (3 in) wide-mouthed, trumpet-shaped, red, pink, violet, purple, white, marbled, velvet-textured flowers are natives of Brazil, and will only do well if grown in pots in the greenhouse. Some gardeners grow them in the bush-house in warm, tropical areas. (Plate 5)

Specialist growers will supply young seedlings, and should you have a heated greenhouse, you can grow them from then on. Or, using the heated greenhouse, you can grow gloxinias from seed sown late winter–spring, provided your greenhouse can maintain a 21°C (70°F) bottom heat.

Sow the seed in sterilised, gloxinia seed compost in seedboxes, pans or punnets, at a 21°C (70°F) temperature. There are many gorgeous, flaunting colours to choose from. The seed is minute and expensive, so mix it in the packet with just a little dry, sharp sand to assist sowing. The seed is pressed into the compost surface gently. Do not bury the seed as this will stifle it. Gently water the seeds and seedlings, keeping the compost pleasantly moist. Do not overwater.

Keep the young seedlings carefully shaded from the hot sun, but don't let them become yellow and spindly. Prick out the seedlings 25–75 mm (1–3 in) apart, into boxes or pans, when they are large enough to handle easily. Let them grow on, and then prick them out into their 100–150 mm (4–6 in) pots.

Plant gloxinia tubers August–November (late winter–spring). Use readily available gloxinia compost in which to grow the tubers. Crock the bottom of the pots well to ensure free drainage. Place the tuber just at compost level, but leave the compost 25 mm (1 in) below the pot top for ease of watering. Grow in a greenhouse. You may be able to grow them in the shadehouse in the warmer parts, although they suffer badly from inclement weather, and look scruffy for much of their life. Gloxinia foliage is easily spoiled by tap or rain water.

Gloxinias need a warm, mottle-shaded, humid spot, and when growing need to be fed with a suitable gloxinia liquid fertiliser. Reduce the watering when the leaves begin to yellow and decay in autumn, and eventually stop altogether. Remove the tubers from the pots when they are dormant. Check for disease and insect pests. Store only the good, sound tubers in dry sand or vermiculite, or similar, in a frost-protected place. Check regularly.

(Sinningia: W. Sinning, Head Gardener, University of Bonn. *Family:* Gesneriaceae)

Grape Hyacinth (see Muscari)

Gymnospermium G. albertii is 120–200 mm (6–8 in) tall, and has 20 mm (¾ in) diameter gladiolus-like corms. It is a cool-climate plant. It has modest 10 mm (½ in), mimosa yellow, star-flowers, with chunky, narrow, 20 mm (¾ in) long, finger-like leaves, borne close around the neck of the flower like a ruff. It is like the winter aconite, but not as grand. It is early flowering, and should be grown as a collector's item, planted 50 mm (2 in) deep, and 75 mm (3 in) apart; or in small pots. Use a sandy, free-draining compost, and keep the plant dry during its dormant summer period.

(Gymnospermium: 'gymno', naked; 'spermium', seed. *Family:* Podophyllaceae)

Gynandriris G. sisyrynchium (syn. Helixyra sis. Iris sis.), a Mediterranean–Asia Minor bulb with its early-flowering, pale mauve, yellow-white centre-spot, iris-like flowers, borne on 100–500 mm (4–20 in) stems, has the unusual habit of the flowers opening early in the afternoon and fading by night. The floppy, grassy foliage is longer than the flowers. It is grown from corms, in pots as a collector's plant using free-draining

Habranthus robustus.

compost. G. *setifolia* is considered a weed in some states. Plant gynandririses in a sunny site, and a place where they can be easily controlled, and keep them dry during their summer dormant period. Destroy any surplus bulbs.

(Gynandriris: 'gynandros', bisexual; stamens-style united. *Family:* Iridaceae)

Habranthus (zephyranthes; vain lily; zephyr lily) Closely related to the hippeastrum, and needing similar treatment, this South American bulb genus is subdivided into twelve species. The 75 mm (3 in) diameter and 100 mm (4 in) long, deeply cut trumpet flower cluster appears on the stem in March–April, depending on species. Stem heights can vary from 150–375 mm (6–15 in). The flower throats are marzipan pink and green with bright yellow stamens, or dull red, or pink with purple base, or pinkish white with yellow tips, or yellow with outside pink-purple stain.

Plant the bulbs 75–100 mm (3–4 in) apart during the dormant season, after flowering, autumn–winter in a free-draining soil, in groups in beds or in the rock garden. The narrow 300 mm x 50 mm (12 in x 2 in), strap-like leaves tend to bend in homage to the ground. H. *robustus* (**plate 6**) has pink, purplish rose, green-based trumpets, and is widely grown. H. *cardinalis* has Tudor-brick red trumpets and yellow stamens. H. *andersonii* (*texanus*) has 30 mm (1¼ in) yellow, copper, crocus-like trumpets with purplish pink stains on the outside, or pink and white with yellow tips. H. *brachyandrus* has a pink trumpet, with purple-black base.

(Habranthus: 'habros', delicate; 'anthos', flower. *Family:* Amaryllidaceae)

Haemanthus (blood lily) Blood lilies, so called, some say, because of the blood-wounds on the white-skinned bulbs or, as is more likely, for their blood-red flowers, are found growing from tropical Africa down to South Africa. The flowers of some species appear as 150 mm (6 in) red goblets filled to overflowing with golden or scarlet stamens. They are followed by the most enormous elephant-ear, 600 mm x 150 mm (24 in x 6 in), leaves. They survive in mild, temperate climates.

Blood lilies prefer a rich, well-drained soil. They are gross feeders and appreciate a dressing of well-rotted compost each year. Plant the bulbs, if they are dormant, in autumn (some recommend spring) before growth commences. Other species are planted during their dormancy period. Place the bulbs just below soil surface. In warmer climates plant them 50 mm (2 in) to 75 mm (3 in) below the surface, 300 mm (12 in) apart. They prefer filtered shade, for example, being protected from the hot, afternoon sun.

The elephant-ear leaves, on many species, will die down in autumn; however, on some species the leaves will remain during winter and spring and die down in summer. In spring or, depending on species, autumn, thick, 150–200 mm (6–8 in) tall (sometimes taller), mottled stalks bearing the flowers, will appear. Increase stock by offset divisions of the mature bulbs during the dormant season.

H. *coccineus*, the paint-brush lily or Scarlet April Fool, is a 150 mm (6 in) tall, scarlet beauty. The blooms appear immediately before the leaves. Its two 450 mm x 150 mm (18 in x 6 in) blunt-pointed, leathery, strap-like leaves hug the ground and look unusual; perhaps 'interesting' is a better word, but to me they look vulgar. Plant the bulbs during the dormant season.

H. *katherinae*, which blooms earlier than H. *coccineus*, has a 600 mm (24 in) stem and numerous, six-petalled flowers, in a 150 mm (6 in) wide, bright red, spider-leg cluster. It has three to six short-stalked, semi-evergreen leaves which grow and develop with the blossom.

H. *magnificus*, the 'Royal Paint Brush', spring–early summer, bears bright scarlet, or occasionally pink flowers, 25 mm (1 in) long, in a dense, rounded head. The 25 mm tall, centre stamens are scarlet-gold tipped. The six to eight, 300 mm (12 in) long, wavy, bright-green leaves, which narrow to a short stalk, develop mostly after the flower has appeared. *H. albiflos*, which has characteristics like *H. coccineus*, bears white flowers, and its huge, blunt-point leaves hug the ground. This species is seldom grown by gardeners, although its red and yellow anthers and redcurrant fruiting berries are attractive.

(Haemanthus: 'haem', blood; 'anthos', flower. *Family:* Amaryllidaceae)

Hedychium (false ginger lily; Indian garland lily) A native of South-East Asia, the evening ginger-scented hedychium is similar to the common canna lily in appearance and treatment, and also suffers in frost-prone areas. There are more than thirty species of hedychium, but gardeners mainly grow *H. gardneranum (gardnerianum)* **(plate 79)**, yellow with red stamen filaments; *H. flavum*, yellow with some orange and bright stamen filaments; *H. coccineum*, pink to red with pink stamen filaments; *H. flavescens*, creamy white tinged yellow; *H. coronarium*, white and sweetly scented.

Hedychiums prefer a well-drained soil rich in well-rotted compost or well-rotted farmyard manure. Top-dress with well-rotted farmyard manure, or apply an organic blood and bone fertiliser before the growing season. Water the plant well during the growing season. Plant in partial shade in northern New South Wales and Queensland, and in full sun in the temperate south, although hedychiums seem to scorch a little in Adelaide. They can be grown as pot, tub or container plants using a coarsely textured compost. Stand the pots/tubs in a saucer of water during the growing season. Apply regular feeds of liquid fertiliser as the manufacturer instructs.

The 'bamboo cane' stems, which cluster in dense proportions, can reach 1.2–1.8 m (4–6 ft). Paddle-shaped leaves 375 mm x 50 mm to 600 mm x 150 mm (15 in x 2 in to 24 in x 6 in), depending on species, are borne on each stem. The 150–450 mm (6–18 in) cylindrical, cluster-spike of paddle-shape petals, 100 mm (6 in) across, appear at the top of the stems in late summer.

Cut down the stems once the hedychium has finished flowering. Old rhizome root clumps can be divided to form sections for planting out, which can be done in winter. Plant these rootstocks just below soil surface, 450 mm (18 in) apart. Seed can also be sown. Destroy unwanted rhizomes.

(Hedychium: 'hedys', sweet; 'chion', snow; *H. coronarium*. *Family:* Zingiberaceae)

Heliconia Of the eighty species of the tropical South American, West Indian heliconias, only a handful are available in Australia. They are grown principally for their lobster-claw bracts, the yellow-green, true flowers being insignificant. Some heliconias are spectacular, others such as *H. bicolor (H. angustifolia)* **(plate 80)**, are not, but interesting nonetheless. The smaller *H. humilis*, 1.2 m (48 in), has boiled lobster, 125 mm (5 in) claws, and glossy foilage. *H. bihai* (wild plantain), 1.75–6 m (6–20 ft), is a tall, boiled lobster-claw heliconia. Both are widely grown because of their scarlet bracts, which zigzag like the prows of yachts up opposite sides of the flowering stem. The 900 mm–1.2 m (36–48 in) long, 200–300 mm (8–12 in) wide, oblong, stalked leaves of *H. bihai* are interesting. *H. wagneriana* produces red, green and flushed-red claws. *H. pendula* reaches to 2 m (80 in), and the flower stem weeps downwards. *H. psittacorum*, the parrot heliconia, is 900 mm–3 m (36 in–10 ft) tall, a canna-like plant that has green-yellow, orangey claws. Heliconias are greenhouse plants, and require a winter temperature that does not go below 13°C (55°F). They need shade and sufficient moisture, and will do well in a hot, humid atmosphere. Some species are summered outside, shaded, in subtropical Queensland. Overwatering or water hanging around the roots will rot the rhizomes. The lobster-claw bracts act as water-catchers for the plant and breeding grounds for insects, so do not keep them too close to the house. Propagate heliconias at the end of winter by removing a piece of the root sucker with a young bud or shoot or two shoots attached, similar to dividing a bearded iris. Small sections can be potted-up and planted, just below compost level, in 125–200 mm (5–8 in) pots. Use a rich, free-draining, loamy compost and grow in the greenhouse. Brazilian and Ecuadorean species are available.

(Heliconia: after Mount Helicon, seat of the Muses. *Family:* Musaceae)

Hemerocallis (Haemerocallis; day lily) This temperate East Asian plant produces flowers which live and usually die in one day. They are replaced rapidly by new flowers. Americans have taken this plant to their hearts, although its copper-gold trumpets are their June weeds.

Day lilies are remarkably free from disease and will grow well in most soils. They should be allowed to spread and colonise an area, but once they outgrow their quarters they should be separated, keeping only the healthiest rhizome roots. Destroy those rhizomes not wanted.

This hardy, herbaceous plant produces 600–900 mm (24–36 in) tall, sedge-like foliage. Out from the foliage arise, 600–900 mm (24–36 in) branched and unbranched stems which produce masses of lily-shaped flowers 100–125 mm (4–5 in) across, either as singles or doubles according to species and hybrid. Plant the roots May–September close to the surface, 300–450 mm (12–18 in) apart, and keep them well watered during dry spells. Do not let the seeds set after flowering, as this reduces plant vigour, unless you are cross-pollinating the plants.

The spread of summer flowers can extend for three months, which is a remarkable length of time. Some species are *H. aurantiaca*, orange-yellow with purple flush; *H. aurantiaca major*, the ubiquitous yellow-orange day lily; *H. flava*, lemon yellow; *H. citrina*, fragrant, sulphur to lemon yellow splashed with brown; *H. fulva*, brick red-orange; *H. thunbergii*, with fragrant, apricot flushed-yellow blooms.

They can be obtained as hybrids in ivory, creamy yellow-green, primrose, yellow and pink with white ring, mimosa yellow, buttercup yellow, gold, fawn, tawny, pure pink, orchid pink, pink with breath of lavender, rose pink, salmon pink, ripe peach, burnt orange, orange-red, chartreuse and blackish. A remarkable colour range.

Some favourite cultivars are: 'Bagdad', pale orange; 'Captain Blood', red; 'Crimson Glory', dark, solid red; 'Golden Glow', yellow; 'Honeyrock', creamy fawn; 'Painted Lady', crimson and brown; 'Patricia Fay', pure pink with gold throat; 'Stafford', sun-kissed peach. There are hundreds more. **(Plates 81-4)**

(Hemerocallis: 'hemera', a day; 'kalles', beauty. *Family:* Amaryllidaceae)

Herbertia (Alophia) The South American *Herbertia platensis* has a 50–100 mm (2–4 in) wide triangle of three, flat, rounded-edge, fragile-looking, porcelain blue outside petals, which surround the inner petals like an upturned, three-cornered hat. The three smaller, inner, crested petals, which are poised inwards like striking cobras, are white and yellow-lobed, with purplish and yellow-barred throats that taper to the flower's centre cup. The flowers appear spring, summer and early autumn. The 900 mm–1.2 m (36–48 in) plant needs the same treatment as the *Tigridia* species (Jockeys' Caps), that is a site which receives the sun, a soil that is well drained, but preferably enriched with compost. The 600 mm (24 in) leaves are straight, sword-pointed, ribbed, greyish blue, like bearded iris leaves. Propagation is by separating the corms in late winter.

(Herbertia: Dr W.M. Herbert, botanist. *Family:* Iridaceae)

Hermodactylus (widow iris; mourning iris; snake's head iris) *Hermodactylus tuberosus* is a Middle Eastern native that has a variety of common names, which represent its appearance. The evergreen, blue-green leaves are 500 mm (20 in) long, squarish, winged, 3 mm (1/8 in) wide, softer and more delicate than the bearded iris. The 45 mm (2 in), fragrant, orchid-iris flowers have cooking apple yellow-green standards (top petals), with funeral black, or lurid purple-black falls (lower petals). It creates much interest in the garden as there are few plants with green and black, orchid-like flowers. It grows well in colder climates, its tuberous, bulbous, pronged roots burrowing deep into the soil.

The flowers appear late winter–early spring on 300 mm (12 in) stems. Plant the roots in autumn–winter, 50 mm (2 in) deep, 100 mm (4 in) apart, and allow them to grow for some years before dividing them up into smaller groups.

(Hermodactylus: 'Hermes', god; 'dactylos', finger — root form. *Family:* Iridaceae)

Hesperantha (evening flower) A South African bulb which is similar to the popular sparaxis lilies. The flowers stay closed until afternoon, when gradually the six-petalled, white or yellow star-flowers dazzle as twilight begins. However, it is the drugging, evening fragrance which makes hesperanthas so desirable.

Plant the bulb-looking corms 75–100 mm (3–4 in) apart in February till end of May, in an open soil, for spring flowering. There are two species available: *H. inflexa stanfordiae (stamfordiae)*, mimosa yellow flowers borne on 300 mm (12 in) stems, and *H. buhrii*, cerise pink outer petals, white inner, and yellow anthers borne on 250–300 mm (10–12 in) stems.

(Hesperantha: 'hesperos', evening; 'anthos', flower. *Family:* Iridaceae)

Hippeastrum Flaunting and exotic, the sixty or seventy species of these trumpeted natives of South America have been hybridised to produce an outstanding range of cultivars. The Dutch, in particular, have spent considerable time on improving hippeastrums, and indeed one form is called the Royal Dutch amaryllis, although it is not an amaryllis.

In Australia the species *H. vittatum* **(plates 86-8)**, with its white and red-striped trumpets, is still grown, but it is rapidly being replaced by its more striking hybrids. These hybrids are used in bedding displays in most states.

Hippeastrums are greedy feeders and need a rich, well-drained soil, which has well-rotted farmyard manure incorporated in its make-up. They are planted 150–225 mm (6–9 in) apart, January to April, with the tip of the 75–100 mm (3–4 in) wide bulb just sniffing the air; some experts suggest that two-thirds should be above ground. In Queensland and like areas, the bulbs should be planted 50 mm (2 in) below ground level to help insulate them from the heat. In frost-prone areas the bulbs should be protected from the frost with mulch. Hippeastrums can be fed with liquid manure once the flower bud appears. Continue the feeding until just before the flower breaks opens.

Hippeastrums are best left undisturbed for several years before dividing the clumps, although they can be lifted in autumn once the foliage has withered, and then dried off and stored in a dry, frost-free, airy place. Common forms can be obtained from seed sown in the greenhouse in seed compost. In three years the seed should produce flowering bulbs.

Some increase their bulb stock by slicing up a bulb in segments as one would segment an apple, with each segment being placed in peat moss, vermiculite compost. Practise with one or two bulbs to begin, and analyse the results before committing too many bulbs.

The waxen, trumpet flowers of the hippeastrum, glistening with 'powdered sugar', are huge, measuring 150–225 mm (6–9 in) across, and a clutch of four to six blooms is borne on the 450–750 mm (18–30 in) tall, hollow stems. The colours of some hybrids have been softened to produce velvety textured shades, hues and stripes of white, pink, red, scarlet, crimson and green. They flower August–November before the leaves appear. However, the fleshy, strap, lance-like, low fan of leaves can grow up while the hippeastrums are flowering.

Favourite hybrids are: 'Apple Blossom', pale pink with blush throat; 'Excelsior', orange; 'Mont Blanc', white; 'Wyndham Haywood', blood red. One can imagine all the variations that can be got from a white, pink and red group of plants.

Hippeastrums make superb pot plants, one 75 mm (3 in) diameter bulb to a 150 mm (6 in) pot, or one 100 mm (4 in) diameter bulb, and above, to a 225 mm (9 in) pot. Use a suitable bulb compost, which is readily available in the nurseries. never let the bulbs dry out during the growing season. Gradually withhold the water supply when the foliage has withered. The bulb will then ripen, and many growers turn the pots on their side to stop excess moisture from entering the pots. Green shoot-tips will appear from the bulb once the growing season starts, and it is then time to repot the bulbs in fresh compost and increase the watering.

Some gardeners do grow hippeastrum species or natural hybrids such as *H. elwesii*, orange-red green throat or pale yellow; *H. procerum (Worsleya procera)*, pale, lilac-mauve; *H. x johnsoni* dull scarlet, white stripe and green throat; *H. aulicum*, deep red, green throat; and others.

(Hippeastrum: 'hippeus', a knight; 'astron', a star. *Family*: Amaryllidaceae)

Hosta (plantain lily; funkia) Late spring and summer-flowering hostas are native to China and Japan. The Chinese species seem to be more 'refined', while Japanese hostas have more attractive leaves. Plantain lilies are so called because their leaves resemble the broad-leaved plantain, *Plantago major*, a lawn weed.

The Chinese *Hosta plantaginea* is becoming a popular plant for those who love subtle green, moist-soil plants. The bell-like, six-petalled flowers are borne in clusters at the top of the leafless flower stem. The white, sweetly scented blooms are 75–100 mm (3–4 in) long, with 40–50 mm (1½–2 in) diameter mouths. Its delicate, refined, heart-shaped, ribbed leaves, 225 mm (9 in) long, are a joy. *Hosta plantaginea* var. *grandiflora* has large, corrugated, shiny green, yellow-ribbed leaves, and produces 100–125 mm (4–5 in), scented flowers.

Hostas will grow only in cool, semi-shaded, damp places such as parts of south-east South Australia, Canberra, Victoria and Tasmania, which excludes them from being grown in the open in many parts of Australia. They can be grown in shadehouses, bush-houses and ferneries, in the soil and as pot plants.

Some scented species and varieties issue more fragrance at night, thus making them desirable as spring–summer plants. The flower stalks can vary in height from 200–600 mm (8–24 in), depending on species, variety or hybrid.

Hostas can be divided after the leaf foliage has died in winter, using the fasciculed (bunched), tuberous root offsets. Replant the roots in rich, damp soil, such as peat gardens or bog gardens, or as marginals alongside a pond, or in the shadehouse soil, 300–600 mm (12–24 in) apart. The young foliage will appear in spring.

Hostas can be obtained in various leaf shades from green to yellow-green to blue. Indeed, *H. sieboldiana (glauca)* produces bluish green leaves, and lavender blush flowers, and is popular because of its colouring. Other hostas are: *H. sieboldii (lancifolia albo-marginata)*, brilliant green, white-edged foliage, pale lilac, darker-streak flowers; *H. fortunei*, pale lilac, 40 mm (1½ in) flowers and large, ribbed, heart-shaped leaves; *H. undulata*, green, splashed-white, wavy, margined leaves, and 50 mm (2 in) pale lilac flowers. There are more. **(plates 112-13)**

(Hosta: in honour of Nicolaus Host, famous physician. *Family*: Liliaceae)

Hyacinthella Originating from Turkey and Eastern Europe hyacinthellas, which look like a cross between a bluebell and grape-hyacinth, are attractive, early-flowering plants. The mostly smooth, deep green leaves are from 60 mm (2½ in) to 100 mm (4 in) in height, and 6–10 mm (¼–½ in) wide.

H. lineata has small, 'shy', deep blue, grape-hyacinth clusters on 60 mm (2½ in) stems; *H. siirtensis* has a meagre collection of baby blue blooms on small stems; *H. leucophaea* has a small group of sky blue flowers; *H. dalmatica* is taller and has a lilac blue cluster of flowers. In autumn, plant the bulbs, which can be up to 30 mm (1⅙ in) in diameter, 50 mm (2 in) deep and 50 mm (2 in) apart in a well-drained, open site, that is not hot or too exposed — in fact a soil similar to one you would choose for grape-hyacinths (*Muscari* species). However, because of their small size, I feel hyacinthellas are best grown as pot plants.

(Hyacinthella: little hyacinth. *Family*: Liliaceae)

Hyacinthoides *non-scripta; H. hispanica;* see *Endymion n-s; E. hisp.*

Hyacinthus (hyacinths) Hyacinth cultivars are produced, principally, from the Middle Eastern, southern European, sky-french blue or white *Hyacinthus orientalis*. Cultivars **(plates 114-17)** such as 'L'Innocence', ivory-white; 'Lady Derby', blush pink; 'Jan Bos', rich pink-red; 'Blue Giant', deep blue; 'Delft Blue', porcelain blue; 'Lord Balfour', violet-blue; and so on, are remarkable in that most of these colours have been selected from the one species, and are not found naturally.

Spring-flowering hyacinth cultivars, whose flowers are available in shades and hues of white, yellow, pink, red and blue, are massed at the top of the 250 mm (10 in) leafless flower stem. The flowers issue a clean, rich, sweet fragrance, instantly recognisable.

Hyacinths prefer a deep, well-prepared, well-drained soil rich in humus, such as well-rotted compost or well-rotted cow manure. They do well in Victoria, parts of New South Wales and Tasmania, but will only grow in cool, suitable areas in other states. Do not use fresh farmyard manure around the bulbs, as it rots them.

Hyacinths can be grown in bowls, boxes or sheltered flower beds in small courtyard gardens. They can also be grown outside in the open where, massed in groups of all one cultivar colour, (mixed colours do not look attractive), they form a stately, extravagant floral design.

Hyacinths are planted in March if substantial autumn rains have fallen, but no later than April for spring flowering. If necessary, flood the soil well before planting. Do not plant the bulbs in a hot, dry soil. Plant the bulb with 75 mm (3 in) of soil above the bulbs. In hot sandy soil plant with 100 mm (4 in) of soil above the bulb. Generally, the base of the bulb should be 125 mm (5 in) below ground level. Place, and allow for, an extra 25 mm (1 in) of clean, sharp sand in the bottom of the hole, to help drainage and root production. In heavy soil you can place free-draining compost in the space above the bulb. Space the bulbs 150 mm (6 in) apart. Planting hyacinth bulbs too early or too shallow leads to poor or irregular flowering.

The leaves and flower-bud will poke up just above ground level, and it is then that they are vunerable to damage. Therefore, any weeds which appear should be removed by hand. Some flowering spikes outgrow their strength and will have to be supported by thin bamboo canes.

The foliage will die back after flowering. Once that happens dig up the bulbs and allow them to dry off, on a sieve or similar, as this allows air to circulate around the bulb, in a cool, dry, airy place.

Hyacinths make magnificent pot plants. Use well-moistened (not sodden — squeeze it to find out), bulb fibre compost, which is kept moist. You can grow them singly in pots, and cover the pots with 75 mm (3 in) of granulated peat, or sand, or another clay pot with the drainage hole blocked up to exclude light. The white, bud-peaks will appear. Gradually introduce them into full light, when the buds will turn green. Once you establish height and colour, you can transfer them, grouped, into large bowls for a colourful display. You can also grow hyacinths in pots large enough to allow for the 75 mm (3 in) cover of compost over the top of the bulb.

Forcing

Hyacinths can be forced to flower earlier provided that in the initial stages they can be kept in a coldish temperature of 7°–9°C (44.6°–48°F). They need this cold temperature to form roots. Plant one hyacinth bulb to a 100 mm (4 in) pot in John Innes Compost No 1, and water thoroughly. Place in a frost-free cool area (see above temperatures), in a dark shed or cellar, or place the bulbs outside in a wooden frame and cover with 150 mm (6 in) of well-dampened granulated peat, which must be kept damp. The bulbs must always be kept damp to encourage the roots to form. Constantly check the bulbs to ascertain when the roots have formed, and when the foliage is approximately 50 mm (2 in) tall. Place the bulbs in the home out of the sunlight for about seven to ten days. Then gradually introduce them into light by a window to flower. The light must be correct — not too hot, shining through the windows to scorch the foliage; diffused light is not suitable; high room temperatures will upset the flowering balance.

It is expensive, but good practice, to buy in fresh bulbs each year, as the magnificent, first-year display is not repeated the second or succeeding years.

There are more than fifty hyacinth cultivars to choose from and I have mentioned some favourites, but there are others equally beautiful. Double forms of hyacinths are also available.

The Roman hyacinth (H. orientalis albulus) is a species of hyacinth grown principally because it appears early in the year. The blue or white bell-flowers are arranged loosely along the stem. Not having the tight, bell-cluster of the pot-hyacinth, it is not as attractive in the garden or in pots, although it can be useful as a cut-flower. H. litwinovii has powder blue bells and a black bulb. For H. candicans, summer-flowering hyacinth, see Galtonia candicans.

Hyacinths can be propagated by off-sets taken from the mother bulb, or you could experiment with a couple of bulbs and propagate them as do the bulb growers. Take a firm, healthy hyacinth bulb and, using a sharpened teaspoon, scoop out the basal plate of the bulb to form a 12 mm (½ in) depression inside the bulb. This cuts across the tightly wound, enclosed leaves contained within the bulb.

The bulbs are placed upside-down on clean, untreated wooden slats, in a 20°C (68°F), heated greenhouse in a constant 80 per cent humidity, and grown on. Young bulblets will form at the leaf junctions inside the scooped-out section in four to

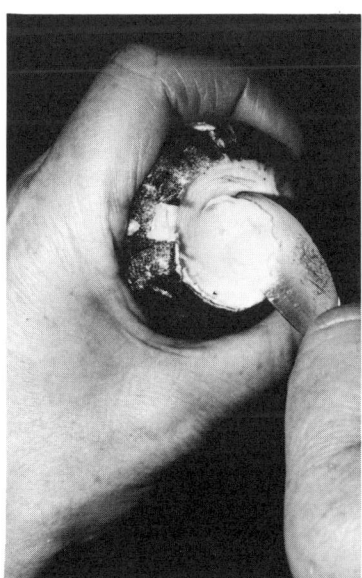

Scooping out base of hyacinth bulb with sharpened spoon to increase bulb production.

six weeks. The temperature is raised to 23°C (75°F), and bulb (with bulblets) is planted in 50 mm (2 in) depth seedboxes or seedpans containing clean, fine, sharp sand. Eventually the bulb will die, leaving many young bulbs to be separated and grown on. This is a complicated process, and much experimenting is required, therefore do not commit too many fine bulbs, greenhouse time and money.
(Hyacinthus: flowers spring from dead Hyakinthos's blood. Family: Liliaceae)

Hymenocallis (Ismenes calathina; Peruvian daffodil; spider lily) The fragile flowers of Hymenocallis narcissiflora (calathina) are similar to a daffodil, but more delicate. Spider lily is a more obvious comparison, particularly when you observe the delicate, curled and tentacled, petal-shreds springing from the white, sweetly drugging, scented, feathery trumpets. The fine flower filaments and anthers attached to the cup, face inwards. The flowers appear in December–January. It is known also as the 'Inca sacred lily.'

The 300–500 mm (12–20 in) tall flower stems emerge from a vast cluster of 600 mm x 50 mm (24 in x 2 in), bright green, strap-like, evergreen or deciduous, depending on species, pointed leaves. From the top of the stem, in mid-summer, erupts a group of three or four, sometimes five, flowers. The trumpets are white, lusciously fragrant, 75–100 mm (3–4 in) long, and 50 mm (2 in) wide, with a greenish flush on the tube.

The bulbs are large like a long-necked onion bulb. They are planted during the dormant winter period in a well-drained soil, 200–300 mm (8–12 in) apart and 125–150 mm (5–6 in) deep. Adding compost to the soil before planting introduces a structure, which is good for the plant. Spider lilies adapt to many situations but prefer a sunny, sheltered site and cool climate. They can be disfigured by frost. Spider lilies are not recommended for north Queensland and similar climatic gardens.

Water the bulbs copiously during dry spells, otherwise the foliage will wilt and once that happens the plants and flowers are never the same. There are over thirty species and hybrids available. H. x festalis (plate 95) is white, elegant and grows well. H 'Advance' has huge white blooms. H. littoralis, a white form, is preferred by some gardeners. There are other white and pale yellow species and hybrids.

Hymenocallises make excellent container bulbs. Ensure that you use a large enough container to accommodate the vigorous bulb growth. Propagate by separating the bulbs in winter.
(Hymenocallis: 'hymen', membrane; 'kallos', beautiful; united stamens. Family: Amaryllidaceae)

Hypoxis (Spiloxene; star flower; star grass, golden weather glass) These six-bladed, aeroplane-propellor flowers are natives of tropical and central Africa, America and Australia. The stems with their prolific flowers are 275 mm (11 in) tall, sometimes even taller and characteristically nod when the fruits form. The centre of the 25–100 mm (1–4 in) diameter flowers are bulls-eyed with brown or black or yellow, depending on the species, and variations within the species (plate 96).

The hollow-stalked H. stellata, 150–175 mm (6–9 in) tall, bears solitary, 18–25 mm (¾–1 in), sometimes much wider, starry, lilac-white, or

yellow or pink, green-keeled flowers. Sometimes none, sometimes three, sometimes six of the petals bear the characteristic purple. Occasionally the petals have an iridescent blue-green, blue-yellow basal blotch. The 300 mm (12 in) leaves can be finely toothed or hairless. The flowers of H. st. elegans are white, with the base blotched purple. H. goetzei has mimosa yellow flowers.

Hypoxis hygrometrica, the 'Australian Golden Weather Glass', is found in wet grasslands and forest clearings, where its fleshy cylindrical tuber soon adapts. The 250 mm (10 in) tall, grassy foliage is sparse and hairy. The flower stalks, which are shorter than the leaves, bear up to three 18–25 mm (¾–1 in) wide, golden yellow flowers. The back of the petals are paler yellow than the front.

Hypoxis foliage can be 210 mm (8½ in) tall, 50–75 mm (2–3 in) broad, and corrugated, often hairy, in varying shades from grey to green to brown.

Don't be surprised if on a cool day, hypoxis flowers stay closed and seemingly praying to the sky, as they only open fully in sunlight. Then the 'brittle-glass' colourful flowers will be totally exposed.

Plant the corms of most species during winter in a sunny position, 100–125 mm (4–5 in) deep and 150 mm (6 in) apart. Separate the corms to increase your stock. Hypoxises make good rock garden plants, but the spot needs to be chosen well to compliment the flowers and foliage. Hypoxises also make fine pot plants.
(Hypoxis: 'hypo', beneath; 'oxys', sharp; seed capsule. Family: Hypoxidaceae-Amaryllidaceae)

Incarvillea (Pride of China) The tuberous-rooted incarvilleas grow naturally in the Himalayas and China. The 'gloxinia' flowers, which appear in spring, are 75 mm (3 in) long, 50–75 mm (2–2¾ in) wide, trumpeted, with purple, geranium-pink with cream or yellow throats. There can be three to eleven flowers on one stem. I. younghusbandi is a dwarf nepalese species, and has glowing pink trumpets with apricot throats. I. delavayi has 300–600 mm (12–24 in) celery-like foliage, carmine pink, yellow-stained trumpets, and 300–900 mm (12–36 in) flower stems. I. lutea has apricot yellow trumpets. Plant the scaly rootstocks summer–autumn 225 mm (9 in) apart with 25 mm (1 in) of soil above the crown. They prefer an open, free-draining soil and a cool climate. Incarvilleas grow freely from seed. A collector's plant.
(Incarvillea: Pierre d'Incarville, Jesuit missionary. Family: Bignoniaceae)

Ipheion (spring starflower) You may see the South American Ipheion uniflora also referred to by such names as brodiaea, milla, triteleia and dichelostemma. However, I. uniflora is its correct name.

Ipheions send a single star-flower from the tip of each slender, 150–225 mm (6–9 in) stem. These flowers can be up to 37.5 mm (1½ in) across, and are star-shaped, six-bladed, like a reflexed (swept-back) propellor, with a distinctive darker stripe down the centre of each petal. Propagate ipheions by bulb offshoots. Plant ipheion bulbs in a sunny, well-drained soil or sandy loam, in autumn, 50–75 mm (2–3 in) deep and 75–100 mm (3–4 in) apart.

The blue-green foliage is attractive, at least until handled, when it gives off an oniony smell. Ipheions do not make ideal cut-flowers. Plant the bulbs in clumps, in a partially shaded spot in a free-draining soil, and allow them to spread before dividing them. They spread rapidly so place them where they can be controlled easily. You can top-dress the beds with well-rotted cow manure during the dormant season. They make splendid pot plants, three bulbs to a 125 mm (5 in) pot, in John Innes No 1 Compost, or similar. Destroy any surplus bulbs.

I. uniflora is the most frequently grown spring starflower, and its whitish blue, star-shaped, soapy, apple-mint scented flowers lie flat over on the foliage, and make a splendid splash of colour. Cultivars are: 'Froyle Mill', pink-purple; 'White Excelsior', white; 'Wisley Blue', silver-lilac; 'Golden Harvest', buttercup yellow.

(Ipheion: origin of name not known. *Family:* Liliaceae)

Iris Iris species can be split up into two definite groups: rhizome-rooted and bulb-rooted. The rhizomes look like large, wrinkled sausages with roots, and the bulbs look like small onions.

The rhizome-rooted irises can be divided into bearded (pogon) iris; beardless (apogon) iris; and crested iris, (Evansia: mostly *I. japonica*).

The bearded iris (plates 97–100) can be subdivided into Tall Bearded — up to 1.2 m (48 in) tall; Intermediate bearded — up to 350 mm (14 in) tall; Dwarf bearded (*I. pumila* x *I. chamaeiris*) — up to 250 mm (10 in) tall.

Bearded Irises

The bearded iris has three upper petals known as 'standards' and three lower petals known as 'falls'. Their common name arises because they have a little hairy fringe-beard at the base of the lower petals (falls).

Until recently the bearded iris was considered a flower leper, mainly tired blue or washed-out white or jaundiced yellow. Previously thought to have been developed from a species known as *I. germanica*, bearded irises have now been credited with parentage from *I. pallida* and *I. variegata*. The Americans, British and Australians have spent time and money perfecting the bearded iris. Today there are such colour combinations as mouth-watering lime and tangy yellow; polished bronzed and brick-reddish; parson black and Oxford blue; sun-kissed apricot and Jersey cream; old bronze and sunset orange; pure white and bright chrome; Jersey cream and wine green chartreuse; flesh pink and old tangerine; and so on.

Soil Preparation To grow good irises you must prepare the ground well in advance. Bearded irises like a neutral or slightly sweet (limy) soil, but will not tolerate a too-acid soil. Refer to section on soil testing and the use of gypsum. Dig over the soil and eradicate weeds which could cause maintenance problems, such as convolvulus (bindweed), kikuyu, couch and similar. Bearded irises need good drainage and grow well in sandy soil. Heavy soil must be broken up using clean, coarse, sharp sand; but not organic material such as fresh compost or fresh farmyard manure. It is not easy or cheap to break up heavy soils using sharp sand only.

Although not generally recommended, you can add old, well-rotted farmyard manure to the soil, or old decomposed compost, and dig it well under, making sure it is below the area where the rhizome roots will rest. Iris roots rot if in contact with manure or fresh compost.

The manure dressing should last for three years, which is long enough, because by then the iris clump will have increased sufficiently to be divided into smaller clumps and replanted. The

farmyard manure dressing can be supplemented by an iris flower fertiliser dressing, usually high in superphosphate, which is applied according to the manufacturer's instructions for bearded irises.

Planting Begin by planting the rhizomes in late spring (November–December) just after flowering. You can, if you have no choice, plant in autumn. Use small, young sections of rhizome, which have two fans of young, healthy leaves, buds and healthy fibrous roots. A rhizome is the fleshy root of the iris. Reduce the fan of leaves by half its length at planting time. This iris root should flower in the first season. The rhizome is planted up to its waist; this means half the depth of the rhizome is buried. Scoop out a trough 100–150 mm (4–6 in) deep. Hold the rhizome against the side of the trough with the top of the rhizome at ground level. This is done in anticipation that when the rhizome is watered-in the soil will settle to about midway down the rhizome. Use your commonsense when planting. Firm the soil around the fibrous roots, taking care when doing so. Plant in clumps of six, the plants being 250 mm (10 in) apart.

Maintenance Irises need regular cleaning. Remove weeds, dead leaves or flowers. Spray for insects when necessary. Water the rhizomes well during dry spells. The plants will produce side leaves, and it is then time to apply a dressing of complete bearded iris fertiliser as the manufacturer instructs. Check the iris leaves constantly and remove any rotting that occurs around the neck. Do not apply fertiliser as winter approaches, as the rhizomes are going into dormancy. Winter is the time to spring-clean the iris bed, removing dead, decaying leaves and weeds.

Iris Terminology Amoena — irises having whole standards and coloured falls. Look at an iris and note that the standards are the top petals, the upstanding petals, and the falls are the under petals, or the drooping or flared petals. Bicolor — standards and falls of different colours. Bitones — irises bearing differing tones, hues and shades of the same colour. For example, light blue and dark blue on the same flower. Blend — when the colours blend in with one another; the principal colour being yellow, even though it may be hard to distinguish. Neglecta — an iris of pale blue standard, but having a deeper blue fall. Some of the modern irises of differing colours are called neglectas; for example, irises with pink standards and rose pink falls. Plicata — this type of iris has a light background, but is stitched or dotted around the edges, and sometimes over the petals. Self — the iris flower being all one colour.

Iris pallida bears delicate yellow blooms with lime dotted falls, and could be of interest to the iris collector.

Beardless Irises

Iris sibirica (Siberian iris): This beardless iris prefers soils more acid and moist than the bearded iris. It follows the bearded iris, flowering late spring, and therefore is popular with iris lovers. *I. sibirica* flowers reach a height of 1.2 m (48 in), and are surrounded by rapier-like foliage. The flowers can be blue, in most hues and shades; also reddish and white. Leave the rhizomes for some years to colonise an area. Plant the fresh rhizomes, in large clumps, 50 mm (2 in) below soil level, 225 mm (9 in) apart, in late winter.

Pacific coast hybrids are hybrids from native irises found on the west coast of the United

States, usually *I. tenax; I. douglasiana; I. innominata, I. purdyi* and others. They prefer a slightly acid, moist soil. They must have good drainage. The deep green foliage, some semi-prostrate, will not tolerate the harsh summers of some states, preferring mottled, filtered sunlight. The 300–600 mm (12–24 in) tall, fragrant, elegant hybrids flower in spring. The colour range contains hues, tints and shades of yellow, gold, bronze, peach, apricot, lavender and lilac-blue. They do not like to be moved or divided. Species are easily grown from seed.

Iris kaempferi is a Japanese iris. It will grow in occasionally boggy, but not permanently waterlogged, areas, and in many moist, normal garden soils. Plant in autumn. It grows to 1 m (40 in) tall, and produces flowers ranging in colour shades and hues from white to pink to purple. An unusual feature of the Japanese iris is that the huge, 125 mm (5 in) flowers lie flat like monster butterflies above the foliage. Plant the rhizomes in an acid soil as lime is definitely taboo for these plants, 300 mm (12 in) apart. Leave Japanese irises to colonise for some years before dividing the roots.

Iris laevigata is planted close to water, and its soft purple standards and falls seem to settle perfectly among the green softness and reflections of the water. it must be grown in an acid soil.

Iris unguicularis (*Iris stylosa*) is popular because it flowers in late winter–early spring, when other flowers are scarce in the garden. The 450–600 mm (18–24 in) tall, evergreen foliage is narrow, strap-like and dominating. The fragrant flowers are a bright, lilac-blue, keeled violet, streaked violet on the white throat, and are borne seemingly on no stems or 100–150 mm (4–6 in) stems. Grows in partial shade. Thin out old leaves after flowering.

Louisiana irises, from North America, range in height from 250 mm (10 in), *I. brevicaulis*, to 900 mm–1.5 m (36–60 in), *I. giganticaerulea. I. giganticaerulea*, is distinctive, having large purple, blue or white blooms, with the stately Dutch iris aspect. The spring-flowering, Louisiana irises are natural varieties and cultivars produced from three species of wild iris found in the Mississippi River basin, or more correctly, the bayou country of Louisiana: *I. fulva, I. brevicaulis* and *I. giganticaerulea*. Some include *I. virginica*, but this is not so. Hybrids are available in white, buff, yellow, pink, bronze, copper, lavender, blue, purple and shades of these colours. They are similar in aspect to a cross between the Dutch iris and the Japanese iris (*I. kaempferi*), having the stateliness of the former and some of the fullness of the latter.

They are moisture-loving plants, often found growing naturally in water in warm climates. Elsewhere, however, the hybrids will thrive in a rich organic soil that is moisture-retentive. They are adaptable and will grow in most garden soils in most states, provided the soil is not limy, as they abhor lime. Give them adequate moisture during their spring–summer growing period, but not during their summer dormancy period. Keep them well mulched to prevent the rhizomes from becoming sunscorched. Ideal soils are those prepared by the addition of acid granulated peat, acid well-rotted compost and acid leaf mould, as these materials have the humic property of absorbing moisture, and holding it, so that it can be taken up by the plants.

Louisiana irises do not like having their rhizomes exposed when planting, as happens with the bearded iris. The rhizomes are planted 25–

50 mm (1–2 in) below soil surface. Place a 75 mm (3 in), layer of organic mulch of acid compost, acid leaf mould, acid granulated peat, or well-decayed hardwood sawdust over the root area when the plants are growing well. Failing this, place a 50 mm (2 in) layer of acid, sandy loam over the rhizomes, which can be removed in autumn when the fear of sun-scorch is over. These mulches will pay dividends in water-retention, plant growth and flower production. Plant the rhizomes 225–450 mm (9–18 in) apart, depending on the vigour of the plant, and leave to multiply. They can be lifted and separated once they have outgrown their allotted area, after about two or three years.

Louisiana irises, like bearded irises, are increased by taking the young, end-sections of the rhizome containing flowering buds. You can take rhizome root cuttings yearly after the flowering period to have a continuity of young plants. Some Lousiana irises are evergreen and they, and the deciduous forms, are subject to frost damage and will not grow in the coldest areas. They need to be planted in sheltered sites. They will grow in partially filtered shade, but need full sun in colder areas.

Spuria irises are vigorous, spring–summer-flowering, Southern European–Asian plants. They are reasonably hardy plants and can be seen growing in many Australian gardens. The fan of narrow, sword-like leaves can reach 600 mm-1 m (24–40 in) in height.

Although colloquially known as Spuria irises, they are not hybrids from *I. spuria*, but mainly hybrids from *I. ochroleuca*, a tall, 900 mm (36 in) species with white, flushed-yellow falls and white-orange standards; *I. monnieri*, 900 mm–1.2 m (36–48 in), fragrant, yellow flowers; *I. aurea (I. crocea)* 900 mm–1.2 m (36–48 in), golden yellow flowers; *I. sintensii (I. urumovii)*, 100–300 mm (4–12 in) stems, bluish purple flowers; and *I. orientalis*, purple flowers. (*I. ochroleuca* and *I. orientalis* are classified as one and the same by some authorities.) But there are those who consider that most, or all, of the above are forms of *I. spuria*. However, it is the hybrids produced from these species that are in demand. Cultivars are available in white, cream, white-yellow, yellow, old gold, light brown, chocolate, blue, purple and violet; and blends, shades and combinations of these colours. Some have deeper veining on the standards, with a large yellow patch on the falls. Some of the newer cultivars such as 'Eleanor Hill', purple with a yellow fall; 'Custom Design', purple-brown with a deep yellow blotched fall; 'Illa Crawford', delicate cream with yellow veining; 'Media Luz', dark blue and yellow; and 'Clark Cosgrove', soft blue, have great potential.

Plant the rhizomes in autumn or spring, about 25 mm (1 in) below surface level. The soil should be neutral, slightly acid or even slightly limy, but do not add lime. It should be well drained but rich in organic material such as well-rotted compost, peat or leaf mould. Water the plants well during the flowering period, provided no significant rain has fallen. They like warm to hot summers to mature the rhizomes after flowering. Excessive watering at this time will rot the rhizomes. The foliage will die down during winter. Fertilise the area during autumn or late winter before growth starts in earnest.

Iris pseudacoris is the 1.5–2 m (60–80 in) tall, flat-topped, 75 mm (3 in), buttercup yellow-flowered iris one finds in wet, marginal areas of lakes, rivers and ponds. It should be grown in like areas.

Crested (Evansia) Irises

Crested irises have a prominent, 'cockscomb' toothed crest at the beginning of the inside of each fall petal, which looks like an eyebrow. *I. japonica, I. wattii* and *I. confusa* are spoken of as being the most distinctive of this group. *I. japonica*, has 300–600 mm (12–24 in) arching, graceful, flower spikes, which produce pale lavender, patterned, orange-crested, frilled, 75–100 mm (3–4 in) orchid-like blooms. It prefers a light, well-drained soil and mottled shade. The leaves are shiny, sword-like and evergreen. There is a variegated form of *I. japonica*. *I. wattii* is now considered to be a hybrid. *I. confusa* is smaller, with almost white flowers.

I. tectorum is the Japanese roof iris, so called because the Japanese thatched their roofs with iris leaves containing the iris rhizomes. *I. tectorum* flowers are lavender blue with snow white crests. It prefers good sunlight and a well-drained soil.

Iris foetidissima, the Gladwyn 'stink' iris, has an irresistible feature, and that is the cluster of brilliant, holly-like berries, which are produced at the top of the 300–600 mm (12–24 in) stems, after the creamy yellow flower has capsuled, ripened and split. It does well in semi-shady areas.

Bulbous Irises

These irises are propagated by bulbs. They are mostly beardless but some, such as *I. boissieri*, do have small beards.

Dwarf bulbous irises are used extensively in rock gardens and similar confined areas. The larger *I. danfordiae* is the earliest to flower, having canary yellow flowers with brown spotting on the falls, borne on 100 mm (4 in) stems. *I. histrio* bears flax blue blooms with mimosa yellow tongue on 150 mm (6 in) stems. *I. h.* var. *aintabensis* is smaller with sky blue standards and falls. *I. histrioides*, 125 mm (5 in) stems, with mauve-blue or pale mauve-lilac flowers, spotted deep violet and mimosa yellow tongue. *I. histrioides* 'Major', 75 mm (3 in) tall, has lovely, blue-violet flowers, with fascinating white spots on the stems. *Iris reticulata* and its cultivars range in height from 75–125 mm (3–5 in), and in colour from light blue to sky blue, from lilac-blue to deep purple-blue, from royal purple to deep royal purple. *I. r.* var. *bakeriana (I. bakeriana)* is 100 mm (4 in) tall, and has a soft, cottonwool blue standard, royal purple fall, spotted yellow. It is fragrant. There are many more miniature irises. All those mentioned above can be used in rock gardens. Plant them 75 mm (3 in) deep, 75–150 mm (3–6 in) apart. They should be allowed to colonise the area once planted.

Xyphium Section

The western Mediterranean bulbous Xyphium Section contains Dutch, Spanish, English and Tangier irises. They are grown extensively as cut-flowers for the home, but they also make superb garden flowers. of the four the Dutch iris stands supreme (plate 101).

The Dutch iris is considered to be a hybrid of *I. tingitana* (the Tangier iris) (plate 102), *I. xiphium* (the Spanish iris) and, most likely, *I. xiphioides* (the so-called English iris). The Dutch have spent considerable time and expertise developing superb hybrids. You should buy the bulbs and grow them for one season, have the cut-flowers and naturalise the bulbs elsewhere in the garden. Buy in fresh bulb stock for the following winter–spring cut-flowers, as fresh bulb stock produces better flowers.

Dutch irises are known colloquially as *Iris* 'Hollandica' to separate them from the other irises mentioned above, and are available in a host of shades and hues: light blue, dark blue, greenish white and buttercup yellow, gold with orange splash, violet-blue and yellow tongue, buttercup standards and lemon falls, arctic white and buttercup yellow falls, and more. Cultivars of great interest are 'Bronze Beauty', bronzy gold; 'Golden Harvest', gold and yellow; 'Wedgewood', summer sky standards, winter sky falls, possibly the most famous; 'Lemon Queen', lemon yellow; 'White Excelsior', pure white with yellow splash; 'White Wedgewood', white with lemon-green falls with buttercup spots on the falls.

Plant Dutch irises in autumn when the bulbs become available; usually February–May, 80–150 mm (3¼–6 in) deep, 100–150 mm (4–6 in) apart.

Iris tingitana, the Tangier iris, produces pale or deep blue, lilac-purple flowers May–June, on long 450–900 mm (18–36 in) stems. The foliage is plain, arching and silver-grey or green but the flowers are superb. The Tangier iris is unpredictable, flowering only when it feels like it. The large, onion-size bulbs have to be cured by a hot sun, therefore plant the bulbs in early autumn, with the tips of the bulbs one and half finger joints below soil surface, and 100–150 mm (4–6 in) apart. The soil has to be well drained and the site exposed to the sun. They will suffer badly in acid soils. You should have a succession of blooms during the spring and on into summer.

Iris xiphioides is called the English iris because it was popularised there from early beginnings. The flower stems are tall, up to 600 mm (24 in). The purplish blue flowers are larger and coarser than the Dutch or Tangier irises and have wide 'wings' on the ends of the falls (lower petals). The falls have a central yellow patch. The English iris is easy to grow, and the colours are variations of purplish blue — never yellow.

Iris xiphium, the Spanish iris, is variable and popular as a cut-flower and garden plant. The flower stems grows to 500 mm (20 in). The colours can vary from violet-blue, with a characteristic yellow or orange blotch in the centre of the falls, to white or yellow. The falls do not have 'wings', thus streamlining the appearance of the flower.

Juno Section

This eastern European iris was no doubt named after Juno, the wife of Jupiter, who was considered to have been beautiful and stately. There is some doubt as to whether the Juno Section are of the *Iris* genus, and they could be a separate genus. What sort of botanical name would you suggest: *Juno reginans-pulcherrima*? Juno, regal and beautiful? That would certainly sum up the flowers of this rarely grown section. *I. warleyensis*, 200–400 mm (8–16 in) tall, with flowers 40–50 mm (1½–2 in) long; standards violet-blue, falls moorish blue with canary-yellow blotch around ridge. *I. cycloglossa*, 560 mm (22½ in) tall, clear lavender blue — lilac falls and standards, with white zone and yellow blotch around the centre-ridge of falls. *I. bucharica (orchiodes)*, 350–450 mm (14–18 in) tall, white standards and falls, or pale yellow standards and falls, or canary yellow standards and falls. There are others.

Irises are a vast, complex family of plants, and you will find books written specifically about them.
(Iris: Greek name given by Theophratus. *Family:* Iridaceae)

Ixia (South African corn lily) The spring-flowering corn lily is a prolific plant. The fragrant, tubular-cup, six-petalled, star-shaped, 65 mm (1¾ in) wide blooms open only in sunlight. A kaleidoscope of colours is available. When topped by racemes of flower buds, the 250-500 mm (10–20 in) tall, graceful, wheat-like stems, look like ears of wheat. The flowers do not last long, but are soon followed by fresh ones.

They will grow well, too well, in most cool, temperate parts of Australia, except Queensland, although they are slightly frost-sensitive. They can be grown as pot plants: six corms to a 125 mm (5 in) pot. They have become a weed pest in certain states. On a dull day there is nothing, but on a hot, sunny day the hills are a blaze of colour. Destroy any surplus bulbs.

From the 250–500 mm (10–20 in) tall, sedge-like foliage springs a host of red, yellow, pink, purple, cream or green flowers on 225–900 mm (9–36 in) stems, depending on species and hybrid. *Ixia viridiflora* is sea-jade green with a red-violet, blackish centre-eye, from which grows beautiful yellow, sometimes purple, anthers. It is a widely grown ixia. *Ixia maculata* (**plate 103**), the spotted corn lily, is also widely grown, producing cream-yellow-pink flowers, with spotted, purple-brown bulls-eye. These will flower from spring on.

There are many ixia hybrids available on the market in all shades of orange, yellow, pink and white, and these are most popular. Some good cultivars are: 'Vulcan', scarlet, tempered orange; 'Uranus' lemon yellow, red centre; 'Blue Bird', white flowers, blue centre; 'Rose Queen', soft pink.

Plant ixias in early autumn, after the foliage has died down, in a well-drained garden loam and sunny site (although they have been known to grow well in woodlands). The depth of planting depends on site and soil conditions, and can vary from 25–100 mm (1–4 in); in heavy soils plant shallow; in sandy soils plant deeper. Space them 75 mm (3 in) apart and leave to spread, but make sure they are contained.

Let the foliage die down once the plant has finished flowering. Do not water the area after flowering as this will prevent the corms from maturing. The hot sun baking the ground above the corm catalyses the energy needed for the next flowering season.

(Ixia: old Greek: bird-lime. Refers to plant's sticky sap. *Family: Iridaceae*)

Ixiolirion 'Ixiolirion' means to look like an ixia. However, this plant comes from a different continent, Middle East–Asia, and from a different plant family, the daffodil family. The bulbs are small, 15–25 in (½–1 in) wide, with giraffe necks. They are planted 75 mm (3 in) deep and 100 mm (4 in) apart in autumn, in a well-drained soil. In mild, frost-prone areas they need a mulch in winter to protect them. In severe frost-prone areas they should be grown in pots and moved into a frost-free area. The bulbs need a hot, dry summer to mature them.

The stiff, 300–450 mm (12–18 in) stem bears a floppy raceme of blue or violet or blush purple, tube-barrelled, 50 mm (2 in), six-petalled, star-shaped flowers. Spring-flowering *Ixiolirion tartaricum* (*montanum*) has 200 mm (8 in), grass-like foliage, and violet star-flowers, with a distinctive, darker stripe down the centre of each petal. They make fine cut-flowers and pot plants. Plant four bulbs to a 125 mm (5 in) pot.

(Ixiolirion: 'ixio', ixia; 'lirion', lily; ixia-like. *Family: Amaryllidaceae*)

Kaempferia The South African and South-East Asian subtropical *Kaempferia* species are frost-sensitive ginger-family plants, and should be grown in the greenhouse. The flowers spring from the ground prior to the leaves, and can be a showy white, yellow, lilac or purple, depending on species. The flowers are followed by the leaves, which can reach 450–900 mm (18–36 in). They make unusual pot plants. The soil should be well drained and enriched with compost. The fleshy rhizomes-thickened tubers, which can be aromatic, are planted in spring at compost surface level or just below. The rhizomes must be kept dry while they are dormant. The plants, like most of the ginger family, need partial shade during the summer and copious watering. Although the *Kaempferia* species are subtropical plants, a heated greenhouse, which can hold a 15.5°C (60°F) temperature in winter, and can be sufficiently cooled in summer, can be used.

The Burmese *K. roscoeana* has white flowers or sometimes lilac-blue with white eye. The two principal leaves, which cover the ground like a large bird protecting its nest, are marked with light green and dark green blotches and zonings, and are almost iridescent. *K. gilbertii*, also from Burma, has 100 mm (4 in) green leaves edged with white similar to hosta foliage. Its 25 mm (1 in) wide flowers are white with a violet-blue striped lip. *K. grandiflora* from Kenya has 60 mm (2¼ in) wide, lavender-blue flowers.

K. aethiopica, *K. natalensis* and *K. decora* are African species; *K. elegans*, with its 100 mm (4 in), tall, 30 mm (1¼ in) thick, scaly, fleshy stem bearing a cluster of frangipani-like flowers, is a South East Asian species.

(Kaempferia: Engelbert Kaempferi, German botanist. *Family: Zingiberaceae*)

Kniphofia (red-hot poker, torch lily) South African kniphofias are well named, as the flaming, scarlet-yellow forms do smoulder like red-hot pokers, whereas the yellow forms glow in dull light like torches.

The evergreen, 450 mm–2.5 m (18–100 in), tufted, rapier-like foliage produces 400 mm–2.5 m (16–100 in) flower stems, bloom February to September depending on species and climate; but mostly are summer-flowering. The 'torch' cone cluster on the scarlet, yellow-tipped cultivar 'Royal Standard' is long, reaching up to 300 mm (12 in). *K. uvaria* (**plate 7**), Maid of Orleans, is a variable species, but usually has orange-red, greenish, yellow-tipped flower torches. The hundred or so hanging-swarm cluster of flowers are small, spaghetti-like to long-bell-shaped, star-ended tubes, with the anthers protruding from the tube. *K. ensifolia* has smaller, greenish white, mimosa yellow flower clusters. The flowers of *K. citrina* are red when in bud, opening to pale yellow. There are many hybrids available.

Kniphofias like a rich well-drained soil, sun and ample moisture when they are growing well. They look superb next to an ornamental pool, where the dazzling spikes are reflected in the water. Leave them to colonise an area for some time and then, if necessary, divide and replant the fleshy roots in spring (September–early October).

(Kniphofia: Johann H. Kniphof, plant illustrator. *Family: Liliaceae*)

Lachenalia (Cape cowslip; soldier-boy; soldiers) Lachenalias look like small, slender, red-hot pokers, or butter yellow hyacinths. These South African bulbs are excellent for pot plant or window box display, particularly the 'Pearson' cross, *Lachenalia x pearsonii*, which is a hybrid of *H. aloides* (**plate 104**) and *L. bulbifera*. It has drooping, waxy, close-packed, 40 mm (1½ in), tubular-bell-shaped, orange-yellow flowers, tipped with sun-kissed orange-red. Lachenalias with this red tipping massed together look as mouthwatering as a marzipan cake decoration.

Lachenalias can have hanging, yellow or pinkish red or purple, waxy, tubular-bell-shaped flowers, borne on 150–300 mm (6–12 in) stems. They are available in many species and hybrids. Those mentioned above are typical. The earliest flowering lachenalias are the species, followed by the hybrids, which appear in profusion in September in the southern states. Plant the 18 mm (¾ in), sphere-shaped bulbs in February–March in a well-drained, fertilised soil, rich in organic matter such as well-rotted farmyard manure or compost, 25 mm (1 in) below soil surface in heavy soil; 75 mm (3 in) below soil surface in sandy soil; and 50 mm (2 in) apart. Restrict watering after the flowers and leaves have died right back, as the bulbs need to be baked by the sun to harden off.

Lachenalias can be grouped or used for border edges, but they must be massed to give effect. They can also be used in rock garden displays. Eventually, they will spread and will need to be separated. When this happens, wait for the foliage to die back, dig up the bulbs, thin them out, and replant the plumpest and healthiest, as these will produce up to three spikes of flowers.

L. bulbifera has coral-red-yellow, pendulous, tubular-bell flowers, with purple tipping. *L. glaucina* is a 'blue' form of Cape cowslip, but can vary from yellow through red, green, lilac and blue to pink. *L. aloides tricolor* is yellow, red-tipped, green-purplish, borne on 300 mm (12 in) stems. *L. quadricolor* is yellow, red-green-purplish. There are many more hybrids. Lachenalias produce attractive 'few-leaved' foliage, which appears in winter. It can be, depending on species and hybrid, 25–50 mm wide, belt-like, blotched, and spreads close to the ground.

Lachenalias make ideal pot plants. Use John Innes Compost No 1 or similar. Water the bulbs moderately until growth begins in earnest. Feed with diluted liquid manure once a week while the growth is strong. Keep the plants watered after flowering, until the leaves begin to die down. Cease fertilising. Gradually dry off the bulbs by decreasing the regularity of the waterings. Stop watering when the foliage has completely died down. Store the bulbs in their pots in a cool, dry, shaded place. Riddle out the bulbs just before growth commences. Keep only the healthiest and plant them in.

Lachenalias are susceptible to mite and aphis attack. Use a suitable insecticide available at your local nursery to combat them. Lachenalias are attacked by fungus disease, particularly in warm, damp locations. Fungus disease can be controlled with a suitable fungicide.

(Lachenalia: Professor Werner de la Chenal, botanist. *Family: Liliaceae*)

Lapeyrousia (Lapeirousia; scarlet freesia) A South African plant that will grow in semi-shade and sunshine, and is more adapted to the cool climates of the southern states. There are many species of lapeyrousia, but the vivid *L. laxa* or *L. cruenta* (**plate 105**) is most commonly grown. You may find it sold under the old name *Anomatheca cruenta*.

During its early stages lapeyrousia looks like a

freesia, but its 37 mm (1½ in) long flowers, which arise from 200 mm x 12 mm (8 in x ½ in) tapered foliage, are grouped, three to twelve, seemingly on one side of the 225 mm (9 in) wiry stems. Imagine the main flower is your head. Now stretch out your right arm. Five or more flowers will shoot up from your arm. They open out into scarlet, tempered carmine, tubular-throated, six-petalled, star-ended, spreading flowers. The throat has three, deeper red insect-sighting blotches at its base.

The bulb-like corms are planted in autumn 50 mm (2 in) below soil level, 75 mm (3 in) apart, and grouped for effect. They can be left to colonise an area, when, if need be, they can be dug up and replanted. They should flower during spring.

(Lapeyrousia: Baron Phillipe de la Peyrouse, botanist. *Family*: Iridaceae)

Leontice The Middle Eastern *Leontice armenaica* is an early-flowering bulb, which is considered a collector's pot plant item. It comes from a hot, dry climate and suffers in damp, cold conditions, and is best kept dry during the summer. The 150–200 mm (6–8 in) solid-looking stem has up to twenty-five single-stalked, mimosa yellow, buttercup-looking, black-eyed flowers sticking out for half its length. The foliage is sage green, and looks like sage foliage. Plant the bulbs singly in pots, in free-draining compost, during the dormant season. *L. leontopetalum* and its sub-species are also grown.

(Leontice: Lion footprint. *Family*: Podophylla-ceae-Berberidaceae)

Leopoldia The Mediterranean, Middle Eastern, early summer-flowering *Leopoldia* species are similar to the *Muscari* species (grape hyacinth) and *Bellevallia* species, and were usually included in with them, although they are a distinct genus of plants. However, leopoldias are not as attractive as grape hyacinths. Indeed they are quite ugly, as the flowers are yellowish, greenish, purplish or brownish, and borne haphazardly either side on stems which can reach up to 600 mm (24 in) in height. They do have their own particular attraction and this makes them ideal collectors' plants. *Leopoldia cosmosa*; *L. c.* var. *monstrosa*, where the sterile, feathery, top flowers appear to have exploded; and *L. longipes*, tight, blue flower cluster when young, are the best known species, and their treatment is as for *Muscari* species (grape hyacinth).

(Leopoldia: not known. Possibly King Leopold. *Family*: Liliaceae)

Leucocoryne (glory of the sun) A Chilean native, *L. ixioides* is the species mostly grown. It has three to nine waxy, 38–50 mm (1½–2 in), star-like, wavy-petalled, azure-lilac, white-centred blooms, which erupt in small clusters from the top of wire-like 300–450 mm (12–18 in) flower scapes (stems). The lilac-blue of the inner flower accentuates the golden anthers found in the throat of the flower. The 300 mm (12 in) long, 5 mm (⅕ in) wide, foliage is grass-like, narrow, linear and limpish.

The leucocoryne prefers a mildly moist site in the garden and likes to be partially shaded. The name 'Glory of the Sun' was given by English plant collector, Clarence Elliott, as the leucocoryne reminded him of the 'Glory of Snow' (chionodoxa) back home.

In nature the leucocoryne is a variable species

with a colour range from white to lilac to dark blue. Some flowers have red throats. Leucocorynes grow massed along the foothills of the Andes, lakes upon lakes of them. They also have a heliotrope fragrance, particularly *L. ixioides* var. *odorata*. However, some experts have classified *L. i. odorata* as a separate species: *L. odorata*.

Plant the bulbs in autumn in a loamy soil, 75–100 mm (3–4 in) below soil level, 100 mm (4 in) apart, and leave them to multiply. This usually takes a long time. They need sufficient water during winter–spring, while growing and flowering, but not during the dormant summer period. The leucocoryne is known for its infrequent flowering — one season a feast, the next a famine. It does not produce many bulbs, but sets seed readily (although this is not necessarily viable). They make wonderful cut-flowers, being deliciously fragrant and lasting well in water. They are also grown as pot plants.

(Leucocoryne: 'leucos', white; 'koryne', club; sterile. *Family*: Amaryllidaceae-Liliaceae)

Leucojum (snowflake) The snowflake is often confused with the snowdrop (*Galanthus* sp.), but it has six segments to the flower, all the same length, in contrast with the snowdrop, which has inner petals smaller than the outer.

Snowflakes come into flower in winter, summer or autumn according to species, and their fragrant white or pink, bell-shaped flowers, sometimes tipped with minute 'green-balls' are a delight. The flowers are borne on 100–450 mm (4–18 in) stems, depending on species and cultivar. *L. vernum*, massed under deciduous trees, makes an attractive groundcover.

Snowflakes can be grown in the garden in the more temperate areas. The damp soil lover, *L. aestivum* **(plate 107)** (summer snowflake), suffers in the hotter areas of northern New South Wales and Queensland, if it is not semi-shaded.

Plant the bulbs of *L. vernum* (spring snowflake) and *L. aestivum* (summer snowflake), February–March with 75 mm (3 in) of soil above the necks of the bulb, and 100–150 mm (4–6 in) apart. They will soon spread, but the closer you plant them the better will be the display. Top-dress the bulb area with deciduous leaf mould if available, provided it is completely broken down. Failing this use acid peat moss. Plant *L. autumnale* **(plate 106)** bulbs with 50 mm (2 in) of soil above the neck, 100 mm (4 in) apart, after the flowers and foliage have died down. However, snowflakes are peculiar as, like snowdrops, they prefer to be moved while in flower, or shortly after, to avoid checking the next season's flowers. But it is not always convenient to do so.

Leucojum vernum the central European spring snowflake, has flower stems 150–350 mm (6–16 in) tall, solitary, fragrant, 12–20 mm (½–¾ in) flowers on a stem. *L. aestivum*, the summer snowflake from Europe and South-East Asia, has smaller, 10 mm (½ in) flowers, between three and seven on one 300 mm (12 in) stem. It will grow in boggy, moist soil conditions. The summer snowflake has a lovely selection-cultivar *L. a.* 'Gravetye Giant', a mass of 25 mm (1 in) blooms, with petals tipped with minute, green, ball-like blotches. This bulb competes successfully with the usual run of summer bulbs. *L. autumnale*, the western Mediterranean autumn snowflake, bears scarce, 10 mm (½ in), white, pink-based flowers on 100–180 mm (4–7 in) fine-wire stems. *L. roseum*, the Corsican snowflake,

has 10 mm (½ in), pinkish blooms, borne on 120 mm (4¾ in) stems.

Many grow the spring snowflake for its early flowers which, although limited on the spathe, are beautiful and welcome.

(Leucojum: 'leukion', Greek name. *Family*: Amaryllidaceae)

Liatris The North American *Liatris* species are easy to grow, and one sees them recommended for poolside planting, among other garden sites. The 600–900 mm (24–36 in) tall, *Liatris spicata*, the Blazing Star or Button Snake-root, is popular and is grown for its summer-flowering spikes. These are long, like slender bottlebrushes, and a mass of small, purplish, lilac-amethyst, or pinkish purple-red, or white, 'thistle-head' flowers, which appear flowering down the stem. These flowers have a luminous quality which gives the plant its common name. The foliage is quite fine, similar to carnation foliage, and is clustered around the stem, but it does increase in size the nearer it gets to the base of the stem. *L. s. albiflora* is a white form. *L. pychnostachya*, 900 mm–1.5 m (36–60 in), has purple flowers.

The tuberous roots are planted in spring, in a well-drained garden soil, to which peat or compost has been added, 300 mm (12 in) apart. Liatrises need copious water when growing well, and must not be allowed to dry right out. The plants die back and become dormant in winter, and then they like a dryish soil. Propagation is by dividing the tangled mass of roots late winter–early spring, or by sowing seed at the end of summer when the seeds have ripened. Varieties and cultivars have been developed.

(Liatris: meaning unknown. *Family*: Asteraceae)

Lilium When is a lily not a lily? When it's a lilium. We make this distinction because so many plants are called lilies. Liliums, although they may differ in appearance from each other, have six petals and six stamens.

Firstly, we shall separate the lilium into its principal growing zones: China, Korea, Japan; Europe; Burma, Nepal, North India; North America; South Russia (Crimea), Turkey.

Liliums can be divided into three groups: trumpet flowers; reflexed petals on drooping flowers; and upright cup-shaped flowers. They are split into sections as follows.

Asiatic Hybrids So called because they have their origins in Asia and have been hybridised to produce many exciting colours, shades and hues, from orange to red to pink to yellow to cream to white. Asiatic hybrids are subdivided into: (a) Early flowering, upright liliums; mostly stem-rooting; (b) Liliums whose flowers face outwards, also mostly stem-rooting; (c) Liliums whose flowers hang down on long flower stalks.

Among these we find the Harlequins, the Patterson Hybrids, Early Tiger Hybrids, Fiesta Strains, Golden Chalice, Enchantment, Valiant, Citronelle. There are many colourful liliums in the above groups.

Martagon Hybrids These are the oriental-looking liliums we call 'turk's cap', where the petals curve back at the tips. The leaves are whorled around the stem like a propellor. They are also known as Martagon-Hansonii hybrids, being a cross between *L. martagon* and *L. hansonii*. They are available in dark red, light red, pink, lavender-pink and yellow. *L. m.* var. *album* is a white variety. **(Plate 112)**

Within this group we find the Backhouse Hybrids, Paisley Hybrids, Marhan and Dahlansoni hybrids. The flowers have small hanging petals, curving back at the tips.

Candidum Hybrids L. candidum (**plate 109**) is also known as the Madonna Lily, as it symbolises purity. It is considered to be the oldest lilium in cultivation. It is a base-rooting, wide, trumpet-mouthed flower. Hybrids have been produced.

American Hybrids These liliums have their origin in America and usually have 'turk's cap' flowers on tall stems. Some are stem-rooting, but many have rhizomatous or stoloniferous roots. L. pardalinum (the leopard lilium), L. canadense, L. parryi, L. humboldtii x occellatum and L. superbum are species of American liliums. They have been hybridised, and the 'Bellingham Hybrids' are well known.

Longiflorum Hybrids L. longiflorum, L. formosanum, L. neilgherrense and L. wallichianum have been used to produce these long-flowered trumpet hybrids.

Trumpet Hybrids Trumpet hybrids have these characteristics: (a) funnel-shaped flowers; (b) bowl-shaped flowers; (c) trumpets hang down; (d) the flowers extend on long trumpets, with mouths that seem to explode like metal bursting open. (Hybrids of L. auratum and L. speciosum are discussed elsewhere.)

Aurelian Hybrids L. henryi crossed with L. regale, L. centifolium and L. sargentiae. Trumpet and Aurelian hybrids could be grouped as one because of their similarities.

Oriental Hybrids Hybrids derived from the most spectacular of liliums: L. auratum and L. speciosum. They are grouped thus: (a) trumpet hybrids; (b) bowl-flowered hybrids; (c) flat-flowered hybrids; (d) hybrids with petals curved back. Of these the Parkman Hybrids have achieved 'glamour' status. 'Lavender Lady' is a great lilium. 'Cover Girl' is an exceptional hybrid, having flowers up to 250 mm (10 in) wide. L. japonicum and L. rubellum have been crossed with L. auratum and L. speciosum to produce other exciting hybrids. (**Plates 119-20**)

Original Species The ninety species are plants that grow naturally in the wild. Hybrids have been produced by crossing these species. Many species are beautiful and are still grown. These include L. speciosum, L. candidum (**plate 109**), L. longiflorum (**plate 111**), L. regale, L. centifolium, L. formosanum and L. wallichanum. They are all species of trumpet lilies.

L. henryi, L. tigrinum, L. davidii (**plate 110**), L. cernum, L. hansonii, L. pumilum, L. lankongense and L. wardii are all reflexed Asiatic species.

L. auratum (**plate 108**) is the most exotic and loveliest species of all. The flowers are sweetly fragrant. However, L. speciosum, with its pink petals and red spots, runs a close second.

L. canadense, L. michiganense, L. pardalinum and L. rubescens are all true American species.

It would not be practical to list all the lilium species. There are Lilium Societies in Australia and New Zealand. Your local Botanic Garden; Parks Department or newspaper may know the address. Check the gardening magazines. (**Plates 108-128**)

What lilium bulbs should you choose to begin with? Here are some fine specimens at a reasonable price, generally available in Australia:

1. Lilium auratum, L. a. platyphyllum, L. a. rubrum
2. Lilium speciosum, L. s. 'Glorious'
3. Parkman Hybrids: 'Cover GIrl', 'Jillian Wallace', 'Lavender Lady'
4. Atomic Hybrids: 'Austral Gem', 'Opera House' (double flower)
5. Trumpet Varieties: 'Snow Angel'; 'Rheingold'
6. Aurelian Hybrids: 'Citronelle', 'Mimosa Star', 'Black Beauty'
7. Asiatic Hybrids: 'Hornback's Gold', 'Nutmegger', 'Wattlebird'

Location The Australian climate is so variable for liliums. They have been called 'Children of the Sun', but this is not necessarily so in many parts of Australia. They will not grow in permanent shade either. Queensland is not considered good lilium territory. Generally, trees providing mottled, distant afternoon shade would be suitable. Many lilium growers provide their own afternoon shade with tall (2 m; 7 ft) paling fencing. Liliums need a cool root run — that is they like their roots to be kept cool. Some liliums are more tolerant of sun than others.

Planting Liliums must have a free-draining soil. A general rule for growing liliums is a porous soil and adequate moisture. It should be slightly acid for certain species, rich in well-rotted compost, but not manure. Manure tends to rot the bulbs. You could dig in acid peat moss and lime-free gravel to open a firm soil. You could, at this stage, add a small dressing of blood and bone fertiliser to the top of the soil, which is lightly raked in and in due course allowed to percolate to the lilium roots. Be careful if you apply artificial fertiliser to the soil as lilium roots object to being in contact with it. There are some liliums that will tolerate lime, such as L. candidum, the madonna Lily, L. henryi and L. pomponium. However, lime is certain death to L. auratum, L. speciosum and others, so be selective. Remember, liliums must have good drainage and a cool root run. That is a rule of paramount importance. However, having said that, there are liliums, such as L. pardalinum, that will tolerate a damp, but not waterlogged, soil.

There are lilium bulbs which produce roots and bulbs on the flowering stem immediately above the mother bulb. These are known as 'stem-rooting' bulb liliums. There are many stem-rooting bulb species and hybrids. It is important to remember this when planting lilium bulbs. Generally, small bulbs should have a 50 mm (2 in) covering of soil above the neck. larger bulbs should have 100 mm (4 in) of soil above the neck. Plant a little deeper in hot sandy soil, and a little shallower in heavy soil. Generally, bulbs should be planted 100–200 mm (4–8 in) deep, stem-rooting species and hybrids being planted the deepest.

Basal-rooting lilium bulbs can be planted 50–100 mm (2–4 in) deep, depending on the size of the bulb. Lilium candidum bulbs should be planted with the top of the bulb just below soil surface level, or 20–25 mm (⁴/₅–1 in), no more, below it.

The soil used for growing the bulbs should be prepared well in advance. A golden rule for growing liliums is that they must be planted as soon as they arrive from the specialist. Therefore, everything — the canes, stakes, labels and planting plans — should be ready.

Scoop out a hole large enough to take a group of bulbs, about five small or three large. Some growers place 50 mm (1 in) of clean, sharp sand over the soil to help drainage. Gently spread out the roots, sloping them downwards slightly, over a

small mound of soil. The very small bulbs can be planted 200 mm (8 in) apart; the small bulbs 300 mm (12 in) apart; and the larger bulbs 450 mm (18 in) apart. Cover with soil. Carefully water the liliums immediately after planting to settle the roots and to force soil gently into every root nook and cranny. Once the weather warms up place a suitable mulch of compost or well-weathered, hardwood shavings over the root area to keep the roots cool. Do not use peat or ashes. Peat draws up water, and ashes can contain concentrations of salt. Liliums must not be allowed to dry out; nor should they be overwatered.

It is essential that you make a plan of what you have planted and where. Label the lilium bulbs immediately and, most important, leave some canes to indicate where you have planted them, as you may damage the young shoots when they appear. Hand-weeding around the liliums is the safest way of removing weeds. The bulbs should be left in one spot for some years to increase, but when they have deteriorated, dig them up and replant the healthiest bulbs immediately.

The taller liliums need to be staked to prevent them flopping. This can be done at planting time, thus avoiding damage to the roots. However, if this is too unsightly you can drive in small stakes of the same diameter at planting time, which can be removed and replaced by larger stakes when the need arises. Tie the lilium carefully, securely, but loosely to the stake. Do not strangle it. The stake, tie and lilium should look as if they belong together.

Liliums can be used in the home as cut-flowers, but do not cut off too many stems, as the lilium needs its green leaves to manufacture the bulb's food. The pollen sacs on the lilium contain much pollen, which is generously given away, particularly all over your best dress or suit as you cut them. Liliums, depending on species and climate, flower from early Christmas to March, but mostly January–February. Remove old, faded and dead flowers and seed pods, but not leaves or stems. The lilium uses much energy producing seed, which is wasted unless you intend to grow liliums using the seed.

The bulbs of the bigger liliums such as L. auratum, the Aurelian hybrids and L. centifolium can be bought 50 mm (2 in) in diameter. L. davidii, L. umbellatum and L. formasanum would not be as large.

Tidying Up Cut down the stalks when they have matured, turned yellow, withered and are dying back. Alternatively, you may be able to remove the stems by placing your feet gently, one either side of the bulb to prevent pulling up the bulbs, and, with your hands close to the ground, giving the stem a quick twist and gently pulling out the stem. If the stem resists this type of operation, cut off the stalks, as on no account should you damage the bulbs unnecessarily. This 'twisting-off' of the stems can leave holes in the soil with easy access for small slugs down to the bulb. Fill these holes. Burn or compost the foliage.

Pot, Tub and Container Work Certain liliums (except many basal-rooting species, but you can include L. candidum) are good for pot, tub or container planting, particularly if you have an alkaline (limy) soil. Grow them in the shadehouse in the same light conditions they would expect outside in the garden — that is, mottled shade-sunlight. Plant them in well-drained, 300 mm (12 in) pots, or larger tubs and containers, in special lilium compost, for example: three parts matured, stacked, friable, acid loam; one part acid leaf mould; one part clean sharp sand. A compost of two parts friable loam; two parts acid peat; one part clean, sharp sand would do provided the soil is sterile and slightly acid. Lilium compost is available from the nurseryman. It is imperative that you use inert (non-poisonous) drainage material, such as broken clay pots, when crocking the bottom of the pots. Label the pots so you know which pot contains what. Repot the bulbs grown in the pots each year, and every other year for bulbs in tubs and containers. This is done usually in late winter before growth begins. Apply a slow-release, granular fertiliser early spring. Pay close attention to watering the bulbs as the pots, tubs and containers will soon dry out.

Pests and Diseases Liliums are attacked by slugs and aphis, and diseases such as botrytis and fusarium. Virus diseases are spread by aphis. The treatment for these problems is covered elsewhere in the book.

Propagation *Bulbil or Aerial Bulbs:* Liliums such as *L. henryi, L. tigrinum, L. sulphureum* and *L. sargentiae* produce small aerial bulbs in the axils of the leaves. You will see the pea-sizes, blackish greenish warts developing on the stem. Place them in seedboxes of compost, once they are ripe and can be dislodged easily from the stem. Grow them on. These bulbils can take up to three years to flower.

Stem Bulbs: Liliums such as *L. auratum, L. concolor, L. dauricum, L. davidii, L. duchartrei, L. formosanum, L. hansonii, L. japonicum, L. longiflorum, L. martagon, L. pomponium, L. regale, L. speciosum, L. superbum, L. tigrinum,* and many other species and hybrids, produce bulbs on the stem at, or just below, soil level. Just how many stem bulbs there will be varies with the species and hybrid of lilium.

The stem bulbs can be removed once the liliums have blossomed and the stem has died. Plant them in their own large pots. Move them into beds of their own during the planting season, away from the parent plant to give them room to develop free from competition.

There is a complicated method of increasing stem bulbs by cutting off the mother bulb during the growing period. This is best left until you become an expert lilium grower.

Stolon Bulbs: Some liliums spread just below soil surface by root-stolons, which is a form of growth similar to couch grass in lawns, but not as prolific. Bulbs form at the end tips of the stolon. These can be removed and planted on. There are many famous liliums that produce stolon bulbs: *L. canadense, L. duchartrei* and *L. superbum,* for example.

Rhizome-Rooted Bulbs: Rhizome roots are similar to stolons except that stolons stay close to, or above, the soil surface, whereas rhizomes travel exclusively underground. The 'Bellingham Hybrids' produce rhizomatous roots, with lilium bulb-crowns arising along the rhizomes. You will observe this when you dig up the bulbs. *L. pardalinum* produces large clumps of roots which can be classified as rhizomatous. Separate the bulb-crowns from the rhizomatous roots, and grow them as separate plants.

Natural Separation There are those liliums, like *L. chalcedonicum, L. pardalinum,* (although the roots give the impression of rhizomes) and the 'Mid-Century Hybrids', which can be divided into two, four or more bulbs depending on the lilium's bulb production.

Scale Propagation To increase your lilium collection using scales, start by lifting the mother bulb after flowering, and after the foliage has died down. Examine the bulb and you will see it is made up of overlapping scales. Carefully prise a few scales from the mother bulb; never too many at one time. You will notice that there is a little bump where the scale was attached to the mother bulb. Dust the breaks on the scales and mother bulb with a suitable fungicide to reduce the incidence of infection. Repot the mother bulb.

Place the scales, points facing up, in acid peat compost. The scales will form bulbs quickly if grown in a temperature of 18.5–23°C (65.5–73.5°F). It is imperative that the bulbs be placed in the compost immediately after they are removed from the mother bulb. If you grow liliums from scales planted in summer, they will grow in beds or boxes without heaters, as the summer temperature can easily reach 18°C. Do not let them scorch.

A method of growing liliums from scales is to partially fill a plastic bag with damp, acid, peat moss. First dip the scales in a suitable fungicide to help prevent fungal attack. Place the scales in the compost. Seal the bag and keep it in a temperature of 20°C (68°F). Watch the scales carefully, and eventually you will see primary roots appearing at the base of the scale, followed by small, scale bulblets. The young scale bulbs are then removed from the bag and grown on in suitable compost.

Liliums from Seed Liliums grown from seed can take from eighteen months to four years to flower, depending on the species. Slow, 'hypogeal' (seed-leaves remaining underground when germinating), germinating seeds need a hot period (summer) to make them germinate; a cold period (winter) to rest and prepare for the next warm period (spring) when they will send up a monocotyledon (single leaf). The leaf appears to be nearly dormant, but it is manufacturing food for the young, underground bulb, and can be easily damaged. This is one good reason for growing seed in seedboxes or trays, as single-leaved bulb seedlings are vunerable to inclement weather, insects and damage. They take up to four years to flower. There are methods whereby this long wait can be circumvented by 'fooling' the seed by refrigeration. This is where belonging to a Lilium Society and talking to the experts can be advantageous.

By contrast the 'epigeal' (seed-leaves carried above ground when germinating) liliums will send up a primary leaf after germination, without benefit of a cold rest period. The true leaves soon follow the primary leaves. They flower unusually in eighteen months from sowing the seed.

Use a compost of equal parts acid peat, clean, sharp sand, and sterilised acid-to-neutral garden loam. If you cannot get sterilised acid-neutral garden loam, do not use unsterlised loam; just increase the percentages of peat and sand. Dust the seed with a suitable fungicide, taking precautions for yourself when doing so. Space the largish seed, 37 mm (1½ in) apart covering the entire seedbox area. Barely cover the seed with a thin layer of compost. Place the box in the shadehouse or greenhouse and keep watered but not waterlogged. Sowing the seed in spring is as good a time as any, although slow-germinating seed can be sown at any time.

Constantly check for fungal attack, particularly 'damping off' disease, and insect attack.
(Lilium: Latin-Greek 'lerion' for the Madonna Lily. *Family:* Liliaceae)

Lily of the Valley (see *Convallaria*)

Liriope The Chinese *Liriope graminifolia* is so called because of its 450 mm (18 in) grass-like foliage, which is a beautiful dark green. *L. muscari* (**plate 129**) is the common, 300–450 mm (12–18 in) tall, 13 m (½ in) wide, leaf form. *L. muscari variegata* is a popular variegated-leaf form. The violet-blue, purplish or white, 55–100 mm (2¼–4 in) long flower clusters look like smaller, grape hyacinths. They are also known as blue-turf lilies. The flowers are followed by small, black berries. These plants are used extensively around buildings as border plants. They like a mottled, shady position and can be clustered under deciduous trees. The flower stems reach a height of 300–400 mm (12–16 in), and make good cut-flowers and pot plants. There are lovely hybrids available: *L. muscari* 'Minor White', *L. m.* 'Blue Spire' and so on.

Plant the tuberish rhizome roots May to September in a free-draining soil, 50 mm (2 in) deep, 100–150 mm (4–6 in) apart. Divide the roots when need be after the plant has flowered. (Liriope: after the nymph Liriope. *Family:* Liliaceae)

Littonia (climbing lily-bells) *Littonia modesta* is a South African plant that likes semi-shade. It will thrive in a free-draining, rich soil under deciduous trees. It also makes a good shadehouse plant.

The fork-shaped tubers should be planted August–September, with 50 mm (2 in) of soil above the crown, 450 mm (18 in) apart, although they will spread. Littonia supports itself by leaf tendrils winding around stakes or canes. This climbing lily has glossy, strap-like leaves, and 1.8 m (6 ft) tall stems. In summer a profusion of open, 37 mm (1½ in), golden orange, bell-shaped, outwards-flaring, six-pointed-petalled flowers will appear. Each flower rises from a leaf junction. Littonia leaves, seemingly, are produced on one side of the plant, and the flowers on the opposite. Littonias make excellent pot plants.

Ample water should be given during the growing season, but witheld once the plant goes into the dormant stage. Littonias can be increased by division of the tubers, taken during the dormant season. They can also be raised from seed, their split seed-cases looking like a row of orange-red peas in a pod. The seeds can take months to germinate.
(Littonia: in honour of Professor S. Litton, botanist. *Family:* Liliaceae)

Lloydia *Lloydia serotina* is an early-flowering and rare species of the lily family, rarely seen in cultivation. It is found naturally in the more remote, cold, mountainous, northern areas of the northern hemisphere. The 100 mm (4 in) grass-like stems emanating from the small (8 mm) grey, onion-skinned bulb, are extremely fine and wiry. The one or two 20 mm (¾ in) star-shaped, white flowers are faintly lined and veined with pink or pink-purple, and have yellow-green centres.
(Lloydia: after Edward Lhuyd (Lloyd), seventeenth century naturalist. *Family:* Liliaceae)

Lycoris (spider lily) Chinese–Japanese lycorises are called spider lilies because the spidery, centre anthers reach way out, up to 75 mm (3 in), and then curve back in, similar to a spider's legs, from the centre of the flower. The flower is funnel-based, opening into six, narrow, recurved flower petals.

Lycorises resemble South African nerines. Indeed, they are called Japanese and Chinese nerines. Lycorises flower at a different time, usually late summer to autumn.

Plant the daffodil-like bulbs in clumps of four to six, 60–100 mm (2½–4 in) deep, in a well-drained soil, mid-summer when they are dormant, that is before they flower or indeed before they begin to bud, 100 mm (4 in) apart. A dressing of blood and bone fertiliser, or slow-acting, low-nitrogen granular fertiliser, can be scattered over the soil once the bulbs are planted. Slowly, it will percolate to the bulb roots. Lycorises may not flower the first year.

Plant lycorises in a site that receives morning sun, not hot afternoon sun. Bulbs relish morning sun and produce better flowers for it. However, they have been grown successfully in light shade. They are frost-prone. The flower buds appear some time before the leaves and are susceptible to damage. You can surround the area with stakes when planting, so that you know where the bulbs are. The 37–75 mm (1½–3 in) flowers are produced on 450–500 mm (18–20 in) scapes (flower spikes), and make ideal cut-flowers with up to four flowers on each stem. However, because of the intermingling of the spider legs they look as though they are part of one huge 150–200 mm (6–8 in) flower. Water the plant copiously, if necessary, once you see the flower buds appear out of the ground.

The narrow, strap-like, 5–17 mm (⅕–¾ in) wide, 300–900 mm (12–36 in) tall, foliage follows the flowers and manufactures the plant food needed for the following year. It is essential to encourage leaf growth. The blood and bone or other fertiliser will be taken up by the roots as it is washed down by rain and irrigation. As with nerines, leave the bulbs to colonise in the same spot for several years.

Lycoris africana aurea, the golden spider lily, with five to eight flowers clustered on one stem, is a splendid apricot-orange-yellow that dazzles against a blue sky. *Lycoris radiata* (**plate 130**) is a near-scarlet spider lily, whose anthers accentuate the 'spider-look' more so than *L. a. aurea*. *Lycoris squamigera*, with its fragrant, rose pink, purple-tinged, splashed-yellow-throated flowers, is also grown. *L. incarnata* has scented, flesh pink flowers. There are others. They make splendid pot plants, but use deep pots, John Innes Compost No 1 or similar, plus a slow-acting, low-nitrogen granular fertiliser.

(Lycoris: the actress-mistress of Marc Antony. *Family*: Amaryllidaceae)

Lysichiton (Lysichitum) The spring-flowering North American 'Skunk Cabbage', *Lysichiton americanum*, is closely allied to the arum-calla lily, zantedeschia, and produces similar 'flowers' or spathes. It is grown as a waterside or bog plant, as it can tolerate these damp conditions easily. It is also used to complement a water-lily pond. The 100–200 mm (4–8 in) 'flower' spathe, which is borne on a 300–450 mm (12–18 in) stem, is canary yellow and folds around the true flower, the spadix, like an open-sided funnel. The 50–125 mm (2–5 in) 'pointing-finger', pale yellow, true-

flower spadix is exposed and sticks up through the centre of the funnel.

Lysichitons likes a rich, organic, bog-marsh soil and will tolerate hard winters, the type it would normally receive. It accepts full or partial sun. The rhizomes can be lifted and separated in late winter. It has an objectional smell, hence its name, but if planted away from the house or leisure area it should not offend. The stemless leaves, which are deep, bright green and paddle-shaped, like the calla lily, follow the flowers, and will dominate that section, therefore give them room to develop. There is a later blooming (3–4 weeks) Japanese species, *L. camtschatcense*, which can have sweetly scented white spathes.

(Lysichiton: 'lysis', loose; 'chiton', cloak; the spathe is cast off after flowering. *Family*: Araceae)

Manfreda The Mexican-South Texan, summer-flowering *Manfreda longiflora* comes from the same family as the tuberose, Agavaceae, although some still include the tuberose in the daffodil family, Amaryllidaceae. It is frost-sensitive and therefore needs to be grown in frost-free areas, or as a pot plant or greenhouse plant in temperate Australia. it is not particularly attractive, having a fleshy, rosette cluster of toothed, 150 mm (6 in) long leaves. The flower stem reaches to 750 mm (30 in) and bears nearly stemless groups of brick red, tubular, open-ended flowers.

M. variegata, the Texan tuberose, has flower stalks that reach from 900 mm (36 in) to 1.2 m (48 in), and bears up to twenty-five fragrant, purplish to white-green, nearly stalkless, narrow, 50 mm (2 in) long, funnel-shaped, open-ended flowers. It has 600 mm (24 in) long, blotched brown leaves, which are 25 mm (1 in) wide and slightly toothed.

Manfredas like a reasonably dry soil and not too much water. They tolerate full sun or a little shade during part of the day. The rhizomes are easily divided and replanted. There are four or five species in cultivation. A bulb for collectors.

(Manfreda: Manfredas de Monte Imperiali, ancient writer. *Family*: Agavaceae or Amaryllidaceae)

Marica gracilis (see *Neomarica gracilis*)

Mastigostyla The early summer-flowering, South American, *Mastigostyla major* is similar in appearance, growth habit and culture to *Tigridia pavonia* (Jockeys' Caps). It needs to be planted in a sunny position, in a well-drained soil enriched with compost, and watered well when growing. The flowers have the same short life span, but are quickly replaced by others. The plant can grow to 500 mm (20 in) in height, and produces flowers that have three larger outer petals, and three inner, in lilac-blue or violet-blue shades.

(Mastigostyla: 'mastigo', flagella; flower style. *Family*: Iridaceae)

Merendera Merenderas originate from Spain, Portugal and central Asia, and can flower in winter, spring or autumn. The blooms of these bulbs look like small crocuses that have been attacked by the cat. They are not crocuses, but are of the lily family, and are more like colchicums. They are attractive, but specialist-grower plants. *M. hissarica*, the Asian merendera, is early-flowering with white, 20 mm (¾ in) blooms on 125 mm (5 in) stems. The leaves are dark green, and look like thick, grass leaves. *Merendera*

pyrenaica flowers in autumn and has pink, 30 mm (1¼ in), shredded blooms borne close to the ground on short stems. Plant the bulbs when dormant 50 mm (2 in) deep and 75 mm (3 in) apart in free-draining soil, or in pots of free-draining compost.

(Merendera: 'quita-meriendas', Spanish for colchicum. *Family*: Liliaceae)

Milla (star of Mexico) The central American *Milla biflora* is often sold, wrongly, as *Brodiaea biflora*. The summer and autumn-flowering *M. biflora* produces six-petalled, fragrant blooms. The six petals, borne on a 12–19 mm (½–¾ in) pale green tube, are white, with a green stripe down the middle of each petal. The flowers lie open and flat to the sky, like a propellor. They can be up to 50–75 mm (2–3 in) wide, in clusters of one, three, six or more, on a 150–450 mm (6–18 in) stem. The 400 mm (16 in) foliage is like fine-leaved grass. The flowers remain open at night, and that is their strength, as the saturating perfume is welcome. They make useful pot plants.

Milla magnifica is a larger, white-flowered, green-striped species, with up to twenty-five flowers on one stem cluster, each flower being fragrant. The foliage is rounded.

Plant the corms when they are dormant 75 mm (3 in) deep and 150 mm (6 in) apart, in a garden border, where they can be controlled, and also where the fragrance can be appreciated.

(Milla: Juliani Milla, eighteenth century Spanish royal gardener. *Family*: Liliaceae)

Montbretia (see *Crocosmia*)

Monstera (ceriman, fruit salad plant) *Monstera deliciosa* (**plate 132**), from central America, is usually grown in large pots or tubs as an indoor plant, but it can be grown in the ground in the shadehouse or glass cool-house, or a moist, shaded garden site in the more temperate, frost-free parts of the country. It is a semi-woody plant, but has to be trained to climb.

The 150–225 mm (6–9 in) creamy fawn flower spathe (spike) is sweetly scented. The monstera has huge, deep green, pale ribbed, glossy leaves, which are holed like a Swiss cheese and deeply cut, lobed and rounded like huge fingers. Aerial roots trail from the jointed stems, and are a worthwhile feature. Propagate by inserting a section of the stem, plus joint, into rich free-draining compost. This plant can reach easily to ceiling height. It is seen as a principal plant in office landscaping, where it is not over-watered, and the leaves are cleaned with water regularly.

(Monstera: monstrosity; alluding to its leaves. *Family*: Araceae)

Moraea (South African iris) The better 'Peacock' forms of moraea produce exotic and exciting 'iris' flowers, although it is pointed out that on certain species, such as *M. spathacea* (spathulata) (**plate 131**), the small 'standards' bunch together at the flower centre to form a tufted claw, which rises above the flat yellow 'fall' petals.

Moraea villosa is a hairy-foliaged, slender, 'iris-flower' plant with flat, 50–75 mm (2–3 in) wide, six-petalled, triangular flowers. The three, 3 mm (⅛ in) wide, inner petals (standards) are thin, narrow and pointed. The three, 25 mm (1 in) wide, outer petals (falls) are rounded, and of fragile appearance. The colour of three outer and inner petals graduates in rings from the centre outwards: white to thin, red stripes to chrome yellow to vivid blue-greenish to deep mauve. The common name of *M. villosa*, 'Peacock Flower', is apt as the petals and eyes have the iridescent sheen of oil on water. (The word iridescence is derived from 'Iridaceae', the iris family).

M. bellendenii (syn *M. pavonia* var. *lutea*) has pale yellow, brown streaked blotches, with soft hair flower-claws. *Moraea bicolor* or *Dietes bicolor* (plate 63) has fleeting 'one-day' flowers, but is much grown in the southern states and in Queensland. It produces rapier-like foliage, and 100 mm (4 in) lemon yellow, brown, peacock-eyed blooms that seem to float on the foliage like butterflies. *M. iridioides* is similar to *M. bicolor*, but the flowers are whitish yellow to pale blue to white, with deeper blue centres. *M. robinsoniana* is the 2 m (7 ft) tall, pure white 'Lord Howe Island Wedding Iris'. *M. spathacea* (*spathulata*) is smaller, 450–600 mm (18–24 in), with 60–75 mm (2½–3 in) wide, mimosa yellow flowers, usually with butter yellow sighting blotches on the 'falls'.

Moraeas prefer a rich well-drained soil, but they are grown in gardens bordering lakes and streams. They require afternoon shade in hot areas. The fibrous-netted corms can be divided in winter, and replanted 75–100 mm (3–4 in) deep and 750 mm (30 in) apart. The grass-like foliage, 1.2–2.4 m (4–8 ft) tall, according to species, appears autumn to spring. The flowers follow, and last fully open from one day in hot climates to three or four days depending on climate and species. The gap is soon filled by other flowers.

Vigorous species of moraea have escaped and become pests in certain states. Plant moraeas where they can be controlled easily, and destroy any surplus bulbs.

(Moraea: after Robert More, English botanist. *Family*: Iridaceae)

Montbretia (see *Crocosmia*)

Musa There are bulbous rhizomatous bananas that are grown commercially, as is evident in Queensland, and those grown principally for ornamental purposes, although both can be used ornamentally, as the large, paddle-shaped, shredded leaves make effective foliage plants. The 'flower' bracts of some *Musa* species are triangular in outline, forming a rose pink propellor. In the centre of these bracts are the true flowers, the fruiting bodies, which even in their early stages have a syrupy deliciousness about them, and are visited by bees. *Musa* species originate from the subtropical and tropical eastern parts of the northern hemisphere. The edible banana, and certain of the ornamental species, will not tolerate the colder climates of the south of Australia.

M. basjoo, the Japanese banana, 5.5 m (18 ft) tall, has rose pink, outside 'flowers' (bracts), large enough to engulf your head. They swan-neck down from the main stem and are cupped like great upside-down shells. The smaller, 3 m (10 ft), rose pink *M. textilis* (plate 133), the manilla-hemp banana, has more delicate rose pink bracts. Both will survive outside in a temperature similar to that of Adelaide, in a warm, sheltered spot, as any fierce winds will tear as the leaves. Even the huge *Ensete ventricosum* (*Musa ensete*) is grown in very large, garden areas in similar conditions. *M. rosaceae*, 900 mm–1.5 m (36–60 in) tall, with 900 mm x 250 mm (36 in x 10 in) leaves and yellow flowers, is one of the finest ornamental bananas.

The smaller ornamental bananas can be grown in large pots and, if necessary, be protected from the inclement winter weather. Species can be propagated from fertile seed or from root-suckers taken in spring. To get true cultivars you would need to obtain grown-on plants from the specialist. Bananas like ample moisture during the summer-flowering period, and a soil that is rich in organic material, such as compost and peat. Ornamental bananas spread freely and will need to be confined to their allotted area. Growing edible bananas is a science, which would be covered in a commercial grower's book. Ornamental bananas need to be tidied up once the shoots have died down.

(Musa: Antonius Musa, ancient Roman physician. *Family*: Musaceae)

Muscari (grape hyacinth) These miniature bluebells cluster like grapes around the stem. However, on closer inspection the 5–7 mm (⅕–⅓ in) bells are like minute, baggy bloomers caught around the leg. They are natives of north Africa, Europe and the Middle East. (Plate 134)

Grape hyacinths are used as edging or border plants, where they should be planted to form a solid mass of blue flowers and deep green foliage. They like a well-drained soil enriched with well-rotted compost or well-rotted farmyard manure, which is dug well into the soil before planting. Plant the bulbs in autumn, 100 mm (4 in) deep and 50–75 mm (2–3 in) apart. Most grape hyacinths flower August–October.

The most popular grape hyacinth is *M. botryoides* 150–300 mm (6–12 in) tall, both the white and pale blue varieties. It grows easily but needs shade in the north. The perfume emitted from *M. moschatum* is as heady as musk. However, it is more noticeably strong when the stems are picked for cut-flowers, as the blooms are only 250 mm (10 in) above ground.

Other species of grape hyacith are *M. cosmosum*, which has sterile, mauve flowers at the top of the spike and brownish blue-green fertile flowers below. *M. neglectum* is fragrant but less floriferous than others. It has light blue flowers at the top of the spike, with navy blue, white-edged flowers below. *M. armeniacum* bears deep blue or white, spicy, fragrant flowers. *M. commutatum* has blackish purple flowers. *M. cosmosum* var. *monstrosum* appears as if all the flowers have exploded, forming a mass of feathers. *M. latifolium* has a top lilac-blue ring of flowers, which gives it a flat-topped appearance.

Cultivars are available, including 'White Pearls of Spain' and *M. armeniacum*, 'Heavenly Blue'. *M. armeniacum* 'Blue Spike' has large, double, flax blue, fragrant flowers. *M. tubergenianum* 'Oxford and Cambridge' has light blue flowers at the top of the spike and dark blue below.

Grape hyacinths are useful pot plant flowers. Use lachenalias as a foil, and imagine a combination of summer blue against marzipan gold.

Let grape hyacinths colonise, which will not take many seasons. Then in January–March dig them up, divide and replant.

(Muscari: 'moschus', musk; refers to flower perfume. *Family*: Liliaceae)

Narcissus (daffodils; jonquils) Daffodils and narcissuses are usually grouped under the convenient heading of 'daffodils', although many people refer only to the 'Trumpet', 'Large Cup', 'Small Cup' and 'Large Doubles' as daffodils.

Daffodils and narcissuses can be grown in the open in the cooler states and in mottled shade in the warmer areas. The site should be sheltered from strong winds, as these would wreak havoc among the blooms. In the Northern Territory and Queensland the soil is too hot for narcissus bulbs to naturalise, and they have to be replaced each year with fresh bulbs. They can be mulched with well-rotted compost or similar during the summer to help keep them cool.

Narcissuses (daffodils) will flower, depending on species, hybrid, cultivar and climate, thus: June–July, (earliest); August, (early); late August (early-mids); September (mid); late September (late); October (latest). In the most spectacular Division 1 group of Trumpet Daffodils (see below), 'Doe Ross', yellow, is one of the earliest; 'Cider', yellow, is early; 'Ansett', yellow, is early-mid; 'Devon Loch', white, is mid-season; 'Cedric Graham', white, is late; 'Spitzbergen', white petals, yellow trumpet-cup, is one of the latest. You can buy daffodils in various colours and forms such as: yellow, yellow and white, yellow and orange, lemon and green, pink, white, double-petalled, perfumed, swan-necked, multi-flowered and novelty (unusual). There are up to twelve hundred species, varieties and cultivars of daffodils (narcissuses). (Plates 135-53)

Narkissus, the son of Cephissus and Leirope, was so beautiful that he was forbidden by the Greek gods on Olympus ever to see his reflection. However, one day he looked at his reflection in a woodland pool, fell in love with himself, gradually pined away and died. From that spot sprang the flower we call narcissus. Another story is that the plant narcissus contains a body-stiffening substance called 'narke'. The word narcotic is derived from narcissus.

Study a daffodil/narcissus and you will see that it contains a trumpet or cup (corona) in the centre, surrounded by an outside ring of petals or saucer (perianth); a yellow cup in a yellow saucer. We shall classify daffodils, jonquils and narcissuses thus:

Trumpet daffodils One flower per stem. The trumpet-cup equal to or longer than the longest petal.

Large-cup daffodils One flower per stem. The cup length being more than one-third, but shorter than the longest petal.

Small-cup daffodils One flower per stem. Cup length no more than one third of the longest petal length.

Double daffodils Double flowers. The cup and saucer are not obvious or detailed, the petals being mixed, and intermingled.

Triandrus hybrids Nodding flowers, one to six on a stem. Cup is bowl-shaped; the petals usually reflexed (recurved, swept-back), and silky. Flowers white or yellow, of obvious Triandrus blood.

Cyclamineus hybrids Small blooms; straight-sided trumpet or cup. Short stem. Cup is papery, narrow with fringed edge. Petals reflexed (recurved, swept back) like cyclamen petals.

Jonquilla (jonquils) hybrids One to six flowers on a stem. Short cup and flat saucer. Fragrant mostly.

Tazzeta hybrids 400–550 mm (16–22 in) stems. One to eight flowers on one stem. Fragrant. Cup mostly yellow, and much shorter than the rounded, flat petals of the saucer; sometimes crinkled.

Poeticus hybrids One, mostly fragrant, flower on one stem. Mostly glistening, white saucer.

Cup yellow, or green, squat, shallow, with wavy edges trimmed with red.

Split-corona narcissus The cup is split for at least one third of its length.

Narcissus miscellaneous All those narcissuses hybrids of garden origin not covered in the above classifications.

Wild species and varieties These are narcissuses and daffodils one would find growing naturally; not those bred artificially by growers.

The Royal Horticultural Society of Great Britain has carried out much research on the classification of daffodils, jonquils and narcissuses. They have split them up into twelve divisions based on the flower, its character and origin of the plant. This information is available in public libraries and places that carry detailed horticultural books. There are daffodil societies in Australia and New Zealand, which hold annual shows. Your local Botanic Garden or Parks Department, local paper or gardening magazine may have the address.

Certain daffodil bulbs are known as double-nosed (DN). You should use the top-class double-nosed daffodils for an exceptional display. Some of these bulbs are so large that they will cover your open hand.

The soil for narcissuses should be well drained, deeply dug, and have well-rotted compost or peat incorporated in the bottom of the trench. A free-draining sandy loam, rich in humus, is excellent narcissus soil. A hot, dry, sandy soil will retard the bulb's development, although it can be improved by using organic material such as well-rotted compost, well-rotted manure or granulated peat. A heavy, wet soil must have its drainage improved, and this can be done by using the organic materials mentioned above, plus clean, sharp sand.

If the bulbs are to remain in the one spot for a number of years, a dressing of blood and bone fertiliser will increase bulb vigour, as will a dressing of sulphate of potash at the manufacturer's recommended rates for narcissus. For bulbs that are to be in the site for only the first flowering, use one of the newer, low-nitrogen, slow-acting granular fertilisers that are recommended for narcissuses. There are fertilisers compounded specifically for certain soils. Check with your local Department of Agriculture Advisory Service to see if there is a fertiliser for your area, but the fertiliser must be suitable for narcissuses.

Daffodils and narcissuses are always at their best when massed. The bulbs are usually ready in late summer–autumn, and they should be planted March–April. However, if no appreciable rain has fallen in late February–March, flood the ground copiously before planting, getting the water well down. You can imagine what would happen to the bulbs if planted in hot soil. The planting depth of the bulb depends on the type of soil and the size of the bulb: some bulbs are as large as an open hand, while others are as small as grapes. Generally, plant large bulbs, measuring from the shoulder of the bulb, in an upright position, 125-150 mm (5–6 in) deep, medium-sized bulbs 75–125 mm (3–5 in) deep and miniature bulbs 50–75 mm (2–3 in) deep. You can plant shallower in heavy soil. Plant the bulbs 150 mm (6 in) apart in garden beds, and 225 mm (9 in) apart if naturalising the bulbs in an area for two or three or more years.

When naturalising the bulbs in woodland, cast the bulbs to scatter them, and plant them where they fall. You can improve the planting spot by digging in sand and peat when planting. Buy narcissuses sold specifically for naturalising. Let them colonise the area for many years.

Narcissuses must be kept well watered once growth has started, provided no appreciable rain has fallen. Be careful not to damage the flowering spikes when working around them. It is best to hand-weed in borders.

Cut off dead and spent flowers. You can tie up the leaves with rubber bands after flowering to keep them looking tidy, if untidiness offends you. If not, let the foliage flop and die back naturally. It is important that the leaves die down naturally as this supplies the bulb with nourishment. Some gardeners plant summer annuals if the bulbs are going to stay there for some time. It is important not to damage the bulbs when digging around them, and to hand-weed the annuals to prevent injuring the future bulb flower.

Lift the bulbs once the foliage is dead and can be pulled off without effort. Dry them in a cool, airy place and sort through them, rejecting any rotten bulbs. To ensure the best garden display you should buy in fresh, prepared stock each year from a reputable grower.

There are many beautiful daffodil/narcissus cultivars. Among the best are: 'Arctic Gold", canary yellow; 'Beersheba', white saucer and cup; 'Butter and Eggs' double, mimosa yellow; 'Carlton' buttercup yellow trumpet and pale yellow saucer; 'Colours' (Colors), butter yellow saucer and orange cup; 'Farnsfield' pale pink-orange cup, white saucer; 'Kiandra', gold saucer with orange and yellow cup; 'King Alfred', buttercup yellow; 'Landmark', white with orange-yellow cup; 'Nautillus', white saucer and lemon yellow cup; 'Petersamo', white saucer and pale lemon cup; 'Pink Fizz', white saucer and pink, jumbled cup; 'Summer Fiesta', whie saucer, deep pink cup; 'Toreador', orange-yellow cup, white saucer; 'Zip', orange cup, white saucer.

Miniature species and their hybrids are used for rock gardens. *N. bulbocodium*, hoop-petticoat, (Algeria, South-West Europe), variable height and colour, 150 mm (6 in), pale yellow to buttercup yellow; *N. cantabricus monophyllus*, (Morocco, Southern Spain), 200 mm (8 in), white; *N. triandrus*, angel's tears, (Western Europe), 100 mm (4 in), swept-back petals, mimosa yellow, pale yellow, white; *N. cyclamineus*, (north-west Europe), 200 mm (8 in), swept-back petals, mimosa yellow; *N. asturiensis (N. minimus)*, (northern Spain), 125 mm (5 in), canary yellow; *N. minor*, (northern Spain), 200 mm (8 in), canary yellow; *N. hedraeanthus*, hoop-petticoat form, (south-east Spain), 80 mm (3¼ in) curved stem, mimosa yellow; *N. requienii (N. juncifolius)*, rush-leaved jonquil, (south-west Europe), 230 mm (9 in), canary yellow. There are others.

Narcissuses are attacked by virus and bacterial diseases, nematodes, bulb-mite, and narcissus-fly. These pests and diseases reduce flower size and streak the leaves. See section on *Pests and Diseases*. Growers are breeding virus-free strains, and old favourites such as 'Carlton', 'Fortune', 'Magnificence', 'King Alfred' and 'Golden Harvest' are among those being produced.

Narcissuses suitable as pot plants are discussed in the relevant section on bulbs for pots and containers.

(Narcissus: as discussed above. *Family:* Amaryllidaceae)

Nectaroscordum The late spring-flowering *N. siculum* originates from south and south-west Europe. The top of the 360–900 mm (14–36 in), green-straw stem erupts with a 100 mm (4 in) wide, cluster of up to fifteen, 16–25 mm (²/₃–1 in) long, 16–20 mm (²/₃–¾ in) diameter blooms. The green, hanging, bell-shaped, pointed-petal flowers have flushed purple or red throats, edged with white. A collector's item. Plant the bulbs in pots of moisture-holding compost in partial shade. Keep them contained, as they can become invasive.

(Nectaroscordum: 'nectaro', sweet; 'scordon', garlic. *Family:* Liliaceae)

Nelumbo Many are surprised to know that the summer-flowering, sacred lotus, *Nelumbo nucifera* **(plate 154)**, is an Australian native plant. It is also a native plant of south Asia, where it is revered as Buddha's flower. A cluster of fragrant, pale pink, open petals surround the butter yellow, watering-can rose centre, and the whole looks similar to a 225–300 mm (9–12 in) wide, large ornamental poppy or peony. The flowers and leaves rise well out of the water. *Nelumbo nucifera* is reasonably hardy, surviving in areas similar to Adelaide. Indeed, a beautiful *N. nucifera* pool can be seen in Adelaide's Botanic Garden. The North American *Nelumbo lutea (pentapetalum)* has mimosa yellow blooms, and is interesting as it is the largest, 250 mm (10 in) wide, North American native flower. *Nymphaea lotus* and *N. caeruleae* are the 'Sacred Lotuses' of the Nile.

Nelumbo nucifera can be grown as for water-lilies. The rhizomes are planted horizontally in crates, which contain a compost of six parts loamy soil to one part well-rotted cow manure. They are planted below water surface in about 300 mm (12 in) of water. Propagation, usually done when growth is just beginning in spring, is by division of the rhizome roots, which are brittle and should be handled carefully. The leaves of *N. nucifera* are large, 300–600 mm (12–24 in) diameter, and are circular, like crinkly-veined, up-facing brass-band cymbals or bowls. The thick stalk is centred underneath. The leaves die back after flowering and the golden brown, dead foliage is a beautiful feature. The seedheads, which look like large-holed, watering-can roses, eventually fall face down into the pool and float, and thus they distribute their seed.

(Nelumbo: Sri Lankan name. *Family:* Nymphaeaceae)

Neomarica (apostle plant; false-flag iris) Summer-flowering neomaricas are used extensively to complement a water-garden or stream setting. They will take full sun, but seem to do well in partial shade. The soil must be well nourished and moisture-retaining, but at the same time well-drained. Neomaricas need ample moisture during their summer growing period. They are frost-sensitive plants.

Propagation is by dividing the rhizomes between late winter and early summer, and planting them 300 mm (12 in) apart. They can also be grown from seed sown in spring. Neomaricas make good pot plants.

The South American *N. gracilis* **(plate 155)**, has 75–100 mm (3–4 in) wide, iris-like flowers. The three outer white or bluish petals form a triangle, while the three smaller centre petals, which face inwards poised like striking cobras, are purple-blue, brown-barred and splashed. The leathery, sword-like foliage can reach to 750 mm (30 in) in height and 20 mm (¾ in) in width, eventually forming sizable clumps.

Neomarica northiana has 900 mm (36 in) tall,

flowering stems and 600 mm x 50 mm (24 in x 2 in) foliage. The fragrant flowers are 100 mm (4 in) diameter, usually with white outers merging to yellow-purple towards the centre. The centre petals are sky blue, barred with yellow and purple.

Neomarica caerulea is 900 mm–1.5 m (36–60 in) tall, and originates from Brazil and Guinea. The leaves are sword-shaped 600–900 mm (24–36 in) long and 40 mm (1½ in) wide. The 1 m (40 in) flower scapes (stems) produce six, 75–100 mm (3–4 in) wide, summer flowers. The outside petals are arranged in threes like a three-cornered hat. They are dark blue, with a lighter blue stripe down the centre of each petal. The three small, inner, brown-barred petals rise tufted, and are poised like striking cobras. The cup formed in the centre of the flower is banded or cross-barred with brown on a yellow backdrop. The flowers last for a day or so, and therefore are not exceptional as cut-flowers. However, they are soon replaced by new flowers.

(Neomarica: 'neo', new; 'marica', nymph. This plant was originally thought to be of the *Marica* genus. *Family:* Iridaceae)

Nerine Three species of South African nerine widely grown are *N. bowdenii*, named after Athelstan Bowden, an English nineteenth century plant collector; *N. curvifolia* var. *fothergillii major* (syn. *N. fothergillii* 'Major'), and *N. samiensis*, the Guernsey (Channel Island) lily, so called because following a shipwreck nerine bulbs were washed up on the beach and naturalised themselves in the foreshore area.

Nerines grow best in a well-drained, loamy humus, slightly acid soil, in a sunny position. They require watering when they are growing well, provided no appreciable rain has fallen. Any water restriction during dry spells will reduce flower and foliage production. *N. flexuosa* (**plate 156**), pink, and *N. f. alba*, white, evergreen, prefer light shade.

A topdressing, rich in compost, should be applied during the dormant period, just before flowering, followed by a complete bulb fertiliser at the recommended rates. The summer sun has been given time to bake the ground around the bulbs, and now the topdressing and fertiliser will supply the food and retain moisture.

Nerines make ideal pot plants, with their 600 mm (24 in) tall stems, and cluster of flowers, six to twelve on a stem. Plant one bulb, buried only half its depth, into a 100 mm (4 in) pot, or three to a 150 mm (6 in) pot, and leave them for two or three years to crowd together. Use a slightly acid compost, and feed them with a complete liquid fertiliser. You can use them as house plants when they are in bud or flower.

Plant the daffodil-like bulbs late January–early March, 37.5 mm (1½ in) below the soil surface, or with their necks exposed to the air in the cooler climates, 75 mm (3 in) apart. Burying them too deeply will reduce the subsequent flowering and foliage production.

Nerines will produce masses of six, shredded-petalled, pink trumpets. Leave the bulbs to colonise an area for five or more years as they resent being disturbed and react by not flowering. However, if you have not disturbed them and they do not produce flowers, dig them up after the foliage has died down in the deciduous forms, and during the dormant period with the evergreen. Divide them and replant the healthiest. Take care as nerine bulbs injure easily.

The 300–450 mm (12–18 in) foliage, usually follows the flowers, but in some species may develop with the flowers. It is an attractive, glossy green up to 25 mm (1 in) wide, and 300–450 mm (12–18 in) tall. Keep the plant watered well while the foliage is growing, provided no appreciable rain has fallen. The foliage will begin to die back at the end of summer in the deciduous forms. Once this happens you can dispense with watering and tidy up the foliage. The summer sun will bake the bulbs, which is essential to ensure good flower production.

Nerines are propagated either from bulbs, offsets or seeds. Bulbs will flower in one or two seasons, but seed can take up to three years or more to flower.

N. bowdenii produces as many as three to twelve 50 mm–100 mm (2–4 in), split-trumpet blooms on one 300–500 mm (12–24 in) stem, resulting in a 150–200 mm (6–8 in) wide cluster. Each bloom has six narrow, wavy-edged, light pink petals. Each of the petals has a deeper-pink rib running down its middle, which curves under at the tip. Issuing from the split-trumpets are down-pointing styles and stamens, that turn upwards at the ends. It makes an ideal cut-flower. *N. bowdenii* flowers March to May and sometimes into winter. The foliage of this species may stay green close to flowering time. There is a white form, *N. b. alba. Nerine bowdenii* cultivars such as 'Pink Triumph', 'Rosea' and the like are a deeper pink.

N. crispa (*undulata*) is smaller and less hardy, but even so its flowers are beautiful. The deciduous *N. flexuosa* is also small and has fine, shredded, pale pink petals. It is well suited as a rock garden plant. *N. flexuosa alba* is a white form. *N. filifolia* has 150–200 mm (6–8 in) grass-foliage with spidery, rose pink blooms, borne on 250 mm (10 in) stems. It is used a lot as a rock garden plant.

N. samiensis is the most popular, and suitable, nerine for Queensland, but it can be seen growing in many gardens north and south. It produces up to eight, satin-sheened, 37 mm (1½ in) red-pink flowers on 450 mm (18 in) stems with stamens that dangle 65 mm (2½ in) out of the trumpets, *N. curvifolia* var. *fothergillii major* (syn. *N. fothergillii* 'Major') bears orange-scarlet blooms, sprayed with gold-vapour, on 450 mm (18 in) stems, with greenish yellow anthers that dangle out of the trumpet mouth like tongues. Even the foliage is a luch blue-green.

(Nerine: the water nymph Nerine. *Family:* Amaryllidaceae)

Nomocharis This charming, summer-flowering, pasture lily originates from the Himalayas and western China. The nodding flowers, up to six on a stem, are each 50–100 mm (2–4 in) in diameter. They are fleshy and fringed, with the three inner petals overlapping the three wide outer petals, like a flattish to reflexed sheriff's badge. They are spotted and blotched in white, yellow, rose, pink, scarlet and purple. The three to six dark green, 100 mm (4 in) long, lily-like leaves are whorled up and around the 700–900 mm (28–36 in) flower stems at regular intervals, like aeroplane propellors.

Nomocharises are alpine plants and do well in cold, semi-shaded sites, in a deep, well-drained, but moisture-holding, acid soil, rich in leaf mould. They will take three or four years to flower. *N. aperta*, has flowers 75–100 mm (3–4 in) in diameter, and is grown because of its soft pink petals, black centre-eye and blood spots. (There is a dark pink form.) *N. mairie*, with flowers 75–100 mm (3–4 in) wide, has white petals, some may be rose pink, with purple-reddish blotches. The inner petals are narrow and heavily fringed, whereas the outer petals are smooth-edged.

Plant the lilium-like nomocharis bulbs with the top of the bulb 50 mm (2 in) below soil surface level, 225 mm (9 in) apart. Nomocharises produce roots above the bulbs like certain liliums, so make sure these roots are covered with soil. Propagate stock by division or by separating scales from the mother bulb or by seed, which is easy using John Innes Acid Seed Compost or similar. The seedlings are potted on singly, when large enough to handle easily, and planted out into their permanent quarters in the third year. However, it can take up to four years from germination for flowers to appear. Nomocharis bulbs do not like being dug up and separated.

(Nomocharis: 'nomos', pasture; 'charis', grace, charm. *Family:* Liliaceae)

Notholirion (false lily) An unusual bulb resembling a lilium. Generally, on closer inspection the pink, tubular-trumpet, six-petalled, star-ended flower, with its deep pink clustered stigma-style, is just enough to separate it visually from a lilium. Notholirions have been in cultivation for only about fifty years.

These Himalayan-north-east Indian bulbs have been separated into six species. *N. macrophyllum* has 30–50 mm (1¼–2 in) long, 50 mm (2 in) wide, horizontal or slightly hanging, tubular-trumpet, star-ended flowers, staggered two to six in a group, towards the top of the pliable 300–600 mm (12–24 in) stems. The blooms have a rich, lilac-pink base to the petals, deepening to deep rose coral, purple at the tips. The long, narrow, strap-like, grassy, basal leaves, which clutch on the ground in a loose rosette, shorten after flowering. *N. thomsonianum*, has up to sixteen, 40–60 mm (1½–2¼ in) long, 30 mm (1¼ in) wide, sometimes white or deep pink, fragrant, trumpet-funnel flowers, with strongly flared-back, end petals, which are borne horizontally, or slightly rising or, when in bud, slightly drooping, on 400–900 mm (16–36 in) stems. *N. campanulatum* has two to six, 50 mm (2 in) long, deepish pink to red, tipped-green, six-petalled, pendulous, split-bell trumpets, borne on 450 mm (18 in), shiny green, corn-like stems.

Notholirions flower in late spring and early summer. They are difficult to grow. They are not particularly happy in a hot or a too cold and frosty climate. They like a cool, root run, and to be kept dry in summer. Their treatment is similar to the lilium. The bulbs are largish, 50–100 mm (2–4 in) long and 25–38 mm (1–1½ in) wide, and covered with a papery, dark brown, cloak-like skin. They die after flowering (monocarpic), but in the meantime have produced numerous offspring, which can be taken and grown on, to provide future flowers. Plant the bulbs 75 mm (3 in) below soil surface, 225 mm (9 in) apart during the dormant winter season.

They make excellent pot plants when grown in cool conditions. Use a John Innes Acid Compost, acid peat compost, leaf mould compost or lilium compost. Grow as for pot liliums.

(Notholirion: 'nothos', spurious, false; 'Lerion', lily. *Family:* Liliaceae)

Nothoscordum *Nothoscordum inodorum* (syn. *N. fragrans; Allium inodorum*) is a spring–early summer-flowering, South American–West Indian plant that is naturalising in South Australia, New

South Wales, Queensland and Victoria. It looks like an insignificant onion plant (without the onion smell), 300–350 mm (12–14 in) tall, bearing a cluster of five or six, 13 mm (½ in), small trumpet, white, pale yellow base, heliotrope-fragrant, 'lily' flowers, with a flush pink stripe running down the outside of each petal. A collector's plant. Can be grown in pots, in free-draining compost, contained, with surplus bulbs being destroyed.

(Nothoscordum: 'nothos', false; 'scordon', garlic. Family: Liliaceae)

Nymphaea (water lily) and Nymphoides

Nymphphaea cultivars, water lilies (plate 11), are famous for their mouth-watering, exotic appearance. They can be fragrant, used as cut-flowers, although they need treating to stop them from closing up. The water can be kept clean by using control methods. The flowers can vary in width from 25 mm (1 in) to 300 mm (12 in). They can flower in daytime or at night depending on species, so check on this. The beautiful, hardy cultivars will survive winters as severe as those of England. Tropical water lilies will thrive in Queensland and similar frost-free areas, although I do hear of certain species surviving in Melbourne. Check with your local Botanic Garden. There are also pigmy water lilies, which are beautiful, if somewhat diminutive.

Allow 0.5 m² (4 sq ft) of water surface for each plant in a small pool, and go up to 0.75 m² (8 sq ft) in a larger pool. The time of planting is usually May to October, although many wait till spring. The depth of water in the pond should be 450–600 mm (18–24 in), but hardy water lilies will cope with 300 mm (12 in) above the tuber rhizome. However, you can place the new water lilies on the floor of an empty pond, and add water so that the tuber rhizome crowns are covered by say 75 mm (3 in) of water. Then gradually add water as new growth begins until the plants reach their final height. Keep a constant check on evaporation from the pond, and top up if necessary.

Water lilies are usually grown in water-lily crates. Some people plant the tuber rhizomes 150 mm (6 in) deep on the floor of the pond, in a compost of six parts, by volume, heavy loam topsoil, to one part, by volume, well-rotted cow manure. This can be messy and impractical in a small garden pond, but practical in a large, natural pond area, where the compost is enclosed by rocks on the pond floor. Another compost mix is two parts, by volume, rich fibrous loam, to one part, by volume, well-rotted cow manure pats. A third compost mix is two parts heavy loam, to one part well-rotted cow manure pats, plus water-lily fertiliser. The crates are like milk crates, being 300–450 mm (12–18 in) or larger in diameter, with enough depth to hold the big tuber rhizome and 150 mm (6 in) depth of compost.

You can use 28.3 or 56.6 litres (1 or 2 bushels) of compost for the less vigorous species and cultivars, and 113 litres (4 bushels) of compost for the more vigorous species and cultivars, which means you have to have the right size crate. A layer of large, grit-size, sharp sand is placed on top of the compost to prevent the soil particles from floating. On top of this you can secure a layer of sticks or chicken wire to prevent the tuber rhizome from floating. on top of this, if necessary, you can place stones to prevent the fish from burrowing into the compost.

Plant one water lily tuber rhizome to a crate —

they soon spread, and this keeps the flowers distinct. The crates should be neutral in colour so as not to stand out under water. Plant the tuber rhizome horizontally or vertically according to species. Certain Marliac cultivars have the vertical characteristic. Check with the supplier. Water lilies must have full sunshine and not be too close to trees. Water lily foliage must be contained so that the leaves cover only half the pond, leaving the rest of the space for the fish and other aquatic animals. Water lilies are usually lifted after two years, or even one year if practical and necessary, separated and repotted.

You can grow water lilies in a large pond, a small pond and even a half wine-cask. Moulded and precast ponds can be bought from the water-plant specialist.

Hardy water lilies are available in white, pink, yellow, apricot and red shades. The tall, tropical water lilies are available in white, pink, blue and purple shades. The pygmy water lilies (Pygmaea species) come in white, yellow, peach and apricot. Tropical water lilies are noted for their fragrance, and many hardy species and cultivars are fragrant.

Nymphoides are treated the same as water lilies and make interesting plants, particularly our own native N. exigua, with its starry, pale yellow flowers and grass-green, one-cent foliage. However, it should be noted that some nymphoides are not desirable as they can become pest plants. You would do well to contact your local Agricultural Pest Plant Officer. Check with your local council, or State Government section in the phone book for the address.

(Nymphaea: Nymphe; Greek diety. Family: Nymphaeaceae). (Nymphoides: like a water lily. Family: Gentianaceae)

Ophiopogon

The Japanese, summer-flowering lily-turf, Ophiopogon jaburan (plate 159), is often confused with Liriope muscari, but although of the same family they are of different genus. Ophiopogon likes a shady, sheltered site and a rich soil. The matt-clump of 600 mm (24 in) leaves resembles coarse grass, hence the name lily-turf. It is also known as white mondo grass. There are variegated forms, O. j. var. aureus-variegatum and O. j. argento-variegatus, which are worth growing. The white ophiopogon flowers are clustered on the top of the stem, like drooping hyacinth flowers, but they still reach above the foliage. Ophiopogon is considered to be a low-maintenance ground-cover, lawn replacement. However, in practice it is invaded by coarser grasses and weeds.

O. japonicus, the dwarf, blue mondo grass, has a scarcity of blue flowers. The vigorous, tuber stolon roots soon produce the blue-green, turf feature of the plant, which is worth considering as a grouping under trees or as border lumps, where when fully grown it waves like a sea. The 150–250 mm (6–10 in) tall, 3 mm (¼ in) wide, tufted blue-green foliage is similar to lawn ryegrass. O. japonicus is grown in partial or mottled shade, is reasonably drought resistant, and will manage on a poorish soil. However, although the clumps that grow under trees seem to manage, they always respond to a good watering in summer, by looking more ornamental.

Propagation is by dividing the runner tuber roots, in spring, into 'spring-onion' rooted plants, 75–100 mm (3–4 in) apart. O. japonicus grows well in subtropical climates, but will also grow in the more temperate southern states. There are other species of ophiopogon.

(Ophiopogon: 'ophis', serpent; 'pogon', head. Family: Liliaceae)

Orchid

Australian ground or terrestrial orchids are precious. They are included, hopefully, to increase the quantity of these environmentally threatened species. You should buy or obtain the tubers from growers or the enthusiast, and not go out into the bush and dig them up, which is not only illegal and silly, but destroys the future for our children. Growing exotic orchids is covered in other books.

Australian ground orchids can be grown in a suitable compost, such as: (a) seven parts, by bulk, loam soil; two parts, by bulk, clean, washed, 3 mm (⅛ in) grit, sharp sand; three parts, by bulk, granulated German or Irish acid peat. (b) one part, by bulk, leaf mould; four parts, by bulk, clean, sharp, washed sand. (c) three parts, by bulk, loam soil; one part, by bulk, clean, washed, 3 mm (⅛ in) grit, sharp sand; one part, by bulk, granulated German or Irish acid peat. (d) two parts, by bulk, loam soil; two parts by bulk, clean, washed, 3 mm (⅛ in) grit, sharp sand; one part by bulk, granulated German or Irish acid peat. Growers feed native orchids with organic fertilisers such as blood and bone or fish emulsion. Others use well-rotted farmyard manure, which has been diluted in water until it is the same colour as weak tea.

Plant the tubers when dormant in 200–300 mm (8–12 in) orchid squat-pots or ordinary pots. Crock them well using pieces of broken pots or stones laid on the bottom of the pot to facilitate drainage. Ground orchids need a good drink when they are growing well. Let the compost nearly dry out, but remain slightly moist, and then give it a good drink; and so on. Do not over-water ground orchids when they are dormant as this tends to rot the tubers. They will grow in light shade in the shadehouse, or outside in a sheltered spot.

Ground orchids that are easy to grow are: Acianthus species, insect orchids (gnat, mayfly, mosquito); Caladenia species, spider orchids; Caleana species, duck orchids; Chiloglottis species, bird orchids; Corysanthes species, helmet orchids; Diuris species (plate 157), doubletails; Glossodia species (plate 158), waxlip orchids; Microtis species, onion orchids; Prasophyllum species, leek orchids; Pterostylis species, greenhoods; and Thelymitra species, sun-orchids.

Slugs, snails and aphids are among the worst pests (see section on Pests and Diseases).

(Orchid: 'orkhis', testicle shape of tubers. Family: Orchidaceae)

Ornithogalum

There are over one hundred species of ornithogalums found in Europe, Africa and Asia. Four popular species are: O. thyrsoides, O. arabicum (plate 160), O. umbellatum and O. caudatum (plate 161). The flowers are usually white, green-striped, six-petalled, star-shaped, and borne in a straggling or tight cluster up the stem. The foliage is lush, broad to narrow, usually shiny, strap-like, with a whitish centre-stripe on some species.

Ornithogalum thyrsoides, or chincherinchee after the noise created by the rustling, dry stems, is an excellent cut-flower as the cluster of up to twenty-five white, 25–50 mm (1–2 in) wide, cup-shaped, star-ended flowers, borne on 450–700 mm (18–28 in) stems, last for ages in water. You may see chincherinchees for sale; which have been dyed red, blue, yellow and orange. O. thyrsoides is frost-prone. Make sure the plants receive ample water during the spring, provided no rain has fallen, and keep up the watering until the flowering period is over. Do not waterlog the plants. There is a golden yellowish form, *O. t. aureum* (*O. miniatum aureum*), which is used for rock gardens or borders. It will spread, so contain it.

O. arabicum, Arab's eye, is so called because of the small, shiny, black-green 'pebble'ovary in the centre of the fragrant, white-cupped petals. The petals are up-facing, star-shaped and carry large, yellow stamens. Up to five flowers, 15–50 mm (¾–2 in) wide, appear on the 600–750 mm (24–30 in) flower stems.

O. umbellatum, European Star of Bethlehem, beloved by Leonardo da Vinci, has large, 25 mm (1 in), white, star flowers, with green and white reverse, carried on short, 150–300 mm (6–12 in) stems. They look heavenly when the sun is shining. Mass them for effect.

O. caudatum has huge, onion-like bulbs, from which emerge 600–900 mm (24–36 in) flower stems, bearing clusters of fifty or more small, green-centred, white flowers. The sloppy, deep green foliage reaches to 600 mm (24 in) in length. *O. balansae* is a minature with 100 mm (4 in) stems, and up to fifteen, 20 mm (¾ in), pure white inner and brushed, meadow green outer, cup-shaped flowers. *O. nutans*, has large, 25 mm (1 in), bell-shaped, white and green, nodding flowers, which have star-ended, 'semi turk's-cap' flared-back petals. They are borne on 150–450 mm (6–18 in) stems. It is a woodland ornithogalum.

Ornithogalums flower early, mid-spring, depending on species. Buy top-quality bulbs and plant them during autumn in a free-draining soil in an open site. Generally they are frost-prone. In Queensland they will need partial shade. Plant the bulbs 125 mm (5 in) deep for the large bulbs, 75 mm (3 in) deep for the medium bulbs, and 37.5 mm (1½ in) deep for the smaller bulbs. The tops of the bulbs should be just below soil surface level. The bulbs of *O. caudatum* will eventually grow above the ground, like large onions. Space the bulbs 150–300 mm (6–12 in) apart in autumn. Ornithogalums are propagated from off-set bulblets or from seed. Keep them contained in your garden, and destroy any surplus bulbs.

Ornithogalums make useful pot plants. Use deep pots and John Innes Compost No 1, or similar free-draining compost.
(Ornithogalum: 'ornis', bird; 'gala', milk. *Family:* Liliaceae)

Oxalis There are over eight hundred species of oxalises, originating mostly from South America, South Africa, Europe and Antarctica. One noted species is *O. pes-caprae*, the soursob, highly prized in Europe for its bitter lemon trumpet, but reviled in South Australia, where in July, August, September and October, it stains the earth in a vast yellow carpet, stretching as far as the eye can see. Although it can harm stock, it was introduced many years ago as stock food.

It is a pity that these beautiful shamrock-like oxalises, in their various colours of white, pink, red, blue and yellow, can get out of hand. However, some species are not as rampant as others, and if you intend growing oxalises you should choose these, grow them in pots to contain them, and destroy any surplus bulbs. Contact your local Botanic Garden, Home Gardens Advisory Service or Pest Plant Office for advice on suitable oxalis species for your area. **(Plate 9)**

(Oxalis: 'oxys', acid; sap tastes acid. *Family:* Oxyalidaceae)

Pamianthe This has to be a botanical nomenclature joke-name — Pam's flower. Named after Major Albert Pam, who first flowered it in Europe, *Pamianthe peruviana*, from north Peru-south Ecuador, is from the daffodil family, and is similar to the *Alstroemeria* and *Pancratium* species. However, pamianthe is distinguished by its numerous winged seeds, and stoloniferous bulbs, which have prolonged scales to form a long 'giraffe' neck (false stem). The bulbs are tunic-coated. Pamianthe is frost-prone.

The leaves, five or more, which erupt from the top of the false stem, are 300 mm (12 in) long, by 16 mm (¾ in) wide. Two to four strongly scented flowers are produced on 500 mm or more (20 in or more) long stems. The flower tube is green, 112 mm (4½ in) long, and slightly curved. There is a yellowish green stripe down the middle of each petal. The white, green-striped, bell-shaped trumpet is 75 mm (3 in) long and 50 mm (2 in) wide at the mouth, the mouth petal-segments being triangular and lobed, with stamens rising from the edge of the segments. Pamianthes are usually spring-flowering.

Plant the 50 mm (2 in) diameter bulbs in a fibrous soil rich in humus. They make good pot plants. They have a short rest period, when watering is restricted, but do not allow the leaves to wither. Water the bulbs well when they are growing well. The treatment is as for *Hymenocallis*. The seeds take twelve to fifteen months to mature.
(Pamianthe: Pam's flower; after Major Albert Pam. *Family:* Amaryllydaceae)

Pancratium (sea lily) *Pancratium illyricum*, *P. foetidum* and *P. maritimum* **(plate 162)** are summer-flowering bulbs from the coast of southern Europe. The white, 50–75 mm (2–3 in) wide, fragrant, papery, trumpeted flowers are clustered in groups of four to twelve at the top of 300–450 mm (12–18 in) stems. They are like white daffodils that have had their trumpets (cups) crimped, and perianths (saucers) shredded. *P. illyricum* has fragrant, 75 mm (3 in) wide, stubby-cupped, white flowers, ten, twelve or more to a cluster. The foliage is fleshy, strap-like, dull, bluish green, 450 mm (18 in) tall, and 20 mm (¾ in) wide. *P. foetidum* has white, 75 mm x 50 mm (3 in x 2 in) trumpet blooms, borne on 600 mm (24 in) stems. *P. maritimum* has up to six, fragrant, papery, white flower-trumpets on a cluster, carried on 300 mm (12 in) stems. It has narrower foliage.

The large, spherical, black-skinned bulbs are planted so that their necks are 75 mm (3 in) below soil surface, 150–225 mm (6–9 in) apart. The bulbs are planted in autumn just below ground level in colder areas. Propagate by bulb off-sets or seed.
(Pancratium: 'pan', all; 'kratos', powerful; medicine. *Family:* Amaryllidaceae)

Paramongaia (cojomaria) The summer-flowering South American *Paramongaia weberbaueri* looks like a cross between a *Hymenocallis narcissiflora*, the Peruvian daffodil, *Pancratium maritimum*, the sea lily or sea daffodil, and the cut-flower daffodil. It has the same papery, flower trumpets as the cut-flower daffodil. The saucer-petals and trumpets are butter yellow and fragrant, with a diameter of 150–175 mm (6–7 in). The flower stems grow to about 600 mm (24 in), and are surrounded by six to seven subordinate, deciduous, grey-green leaves, which are 40 mm (1¾ in) wide, and therefore coarser looking than daffodil foliage. The bulbs are large and daffodil-like. They prefer a well-drained soil. A good pot plant, if kept tightly potted and fed with dilute liquid fertiliser.
(Paramongaia: 'para', near; Mongaia, Peruvian site. *Family:* Amaryllidaceae)

Paris *Paris polyphylla*, a 900 mm–2.5 m (36–100 in) tall, early flowering bulb, comes from north east central Asia, particularly the mountain and forest acid soil areas along the Himalayan line. It has insignificant green outer and yellow inner flowers, with brilliant scarlet seeds. The whorls of 150 mm x 100 mm (6 in x 4 in) large leaves surrounding the stem are glossy green and ribbed like a lawn plantain, and make an attractive foliage plant. Grow in large pots using acid peat compost, in shade. Give *Paris* species the same treatment as *Trillium* species.
(Paris: 'par', equal parts; flower parts regular. *Family:* Liliaceae)

Phaeomeria The summer-flowering Indonesian Torch Ginger, *Phaeomeria magnifica*, a tall, 4.5–6 m (15–20 ft) plant, is grown in Queensland, where the subtropical climate is more in keeping with the plant's natural habitat. The torch ginger produces large cane stems, with long, 450 mm (18 in), pointed leaves arranged alternately up the stem. However, it is the bee-swarm, pink or red or purple-red, 150 mm (6 in) diameter, torch-cone of 'flower' bracts, that is the main attraction. It appears at the top of a 1.5–2 m (5–6 ft) flower stem, like a waxy, plastic-shiny, luminous torch that reflects the light and is similar in flower form to the native New South Wales waratah. Most bracts have a delicate white line tracing around the margin edges of the petals.

Phaeomeria likes a rich soil, mottled shade in summer and copious water when it is growing well. The canes will need to be cut back once they have become scruffy. Propagate by root division, and plant 1.05 m (42 in) apart.
(Phaeomeria: 'phaies', dark (red-purple here); 'meris', parts. *Family:* Zingiberaceae)

Polianthes (tuberose) The fragrance of the Mexican tuberose is sweet and all-embracing, but too cloying for some. Tuberoses make magnificent cut-flowers. The foliage is 500 mm (20 in) long, 5–15 mm (⅕–⅗ in) wide and insignificant.

The tuberose is not a particularly attractive flower, and this is why it exudes such a glorious perfume to attract pollinators. The starry flowers are curved, tubular-shape, with six flaring petals; a double form is widely grown. The blooms are 60 mm (2½ in) long and 25–37 mm (1–1½ in) wide. They are waxy, white to cream-yellow-greenish, sometimes with a breath of pink. They are borne in pairs on opposite sides of the flower stem and cluster in a loose, pointed swarm towards the top of the 600 mm–1 m (24–40 in) spike. The blooms open first at the base of the flower spike and then gradually work up to the top.

Tuberoses grow in most parts of temperate Australia, particularly in the north, but there is a secret in growing them. If you have a tuberose root which has just flowered, and hollow tubes can be seen where the old stems have been removed, you can be sure it will not flower again. It is similar to the reproductive action of the gladilous, with the old corm dying, young corms and numerous spawn being produced. Indeed, many treat tuberoses like gladioluses, in as much as they plant plump, healthy bulbs when the risk of frost is over, and lift the bulbs in due season. To be effective with tuberoses you will need good-sized, well-ripened, ovoid (cigar-shaped), tuberous bulbs.

At the side of the flowered polianthes tuber, or crown, you will see strong, juvenile sidegrowths, like small, pointed cigars, plus tuber-spawn off-sets. It is the strongest sideshoots that flower; not the old tuber or the tuber-spawn.

Polianthes tuberose tubers are planted in late spring, when danger from frost is over, in a spot which receives the sun. Buy in unflowered, tuber-crowns for flower production. Plant the 50 mm (2 in) long, tuber-crown, 50–100 mm (2–4 in) deep and 125 mm (5 in) apart, in a rich, lime-free, well-drained soil. Work well-rotted compost or acid peat, or even well-rotted farmyard manure, deeply into the soil, so it is below the tubers. Firm the soil around the tuber and water it well once the shoots begin to grow. Keep the tuber watered provided no appreciable rain has fallen. Tuberoses can also be grown in pots, using acid peat compost, three to a 150 mm (6 in) pot. They should flower about Easter.

The young tubers can be lifted after flowering, in May, stored in sand. Only the two or three strongest sideshoots are retained, and the clump replanted the following year. The numerous off-sets are stripped off, and the best planted separately in pots or nursery beds to mature. Tuberoses can be left in the ground in a warm sheltered spot for two or three years, when they should be dug up and separated.

There are no species found growing wild, and yet there are eight species in cultivation. Mostly *P. tuberosa* is grown. Hybrids are available, and the double-flowered, 'The Pearl' is an exceptional plant.
(Polianthes: 'polios', shining white; 'anthos', flower. *Family*: Amaryllidaceae-Agavaceae)

Polygonatum (Solomon's seal) The arching branches of *Polygonatum multiflorum* (plate 163), with their green, gold sheen, ribbed, 75 mm x 125 mm (3 in x 5 in) or 100 mm x 150 mm (4 in x 6 in), hovering, winged leaves, bear gooseberry-clusters of tubular, bell-shaped, six-reflexed petalled, white, green-tipped flowers. The 18–37 mm (¾–1½ in) long flowers hang down from the leaf-joint junctions in clusters of two or three. Polygonatum prefers the cooler conditions found in Tasmania, Victoria, New South Wales, parts of South Australia and New Zealand, but can be grown in moist soil, shady sites or shadehouses.

Plant the fleshy, thick, knotty, rhizome roots horizontally in autumn–winter, 50–125 mm (2–5 in) deep, 450 mm (18 in) apart, in a neutral to acid soil. In spring, slender 600 mm–2 m (24–80 in) branches, depending on species, will arch out of the soil, which in turn will be weighed down by clusters of 37.5 mm (1½ in) white flowers. They make useful cut-flowers, lasting for some time in water.

The foliage will die in autumn and can be tidied up. Be careful as the fleshy roots are at surface level. The most popular Solomon's seal is *P. multiflorum*, which also has a double-flowered form and a variegated-leaf form. Others are *P. roseum*, the small, two-flowered, pink form; *P. commutatum* with its 1–2 m (40–80 in) tall stems,

and 25 mm (1 in) yellowish, greenish-white flowers.
(Polygonatum: 'poly', many; 'gonu', joints; jointed rhizome. *Family*: Liliaceae)

Primula *Primula fedtschenkoi* is an alpine, tuberous-rooted, spring-flowering, pink primrose. The 10–25 mm (½–1 in) 'primrose' flowers are borne on rigid, 50–120 mm (2–6 in) stems. The spoon-shaped foliage is low-growing and saw-toothed around its edges. A pot plant collector's item for colder climates, using a neutral–acid, free-draining compost, at one tuber to a pot.
(Primula: 'primus', first. *Family*: Primulaceae)

Prochnyanthes The Mexican, summer-flowering *Prochynyanthes viridescens* is of the same family as the tuberose, that is Agavaceae (although some still place it in the daffodil family, Amaryllidaceae), and prefers similar treatment to it, and to the *Bravoa* species. It also requires a frost-free climate. The 50 mm (2 in) fibre-covered bulbs are planted in spring. The flower stem reaches to 1.5 m (60 in) in height. The 300 mm x 50 mm (12 in x 2 in) lance-like leaves are arranged in loose rosettes around the flower stems. The stem supports 25–30 pairs of 40 mm (1¾ in) long, green-brown, abruptly bent, tubular, bell-shaped flowers. *P. bulliana* is also grown. Pot plants for the collector.
(Prochnyanthes: 'procon', ewer; 'anthos', flower; flower shape. *Family*: Agavaceae)

Pseudobravoa Related to the *Bravoa* and *Polianthes* species, the Mexican *Pseudobravoa densiflora*, a summer-flowering plant, produces a dense spike, more like a loose swarm, of 50 mm (2 in) curved, tubular, yellow flowers, which are borne on a 150–500 mm (6–20 in) tall stem. The bulbs are scaly and fibrous with fleshy roots similar to a lilium. Treat as for *Bravoa* and *polianthes*.
(Pseudobravoa: false bravoa. *Family*: Agavaceae)

Pseudogaltonia *Pseudogaltonia clavata* is the only known species of this close relative to *Galtonia candicans*, the summer-hyacinth. It comes from a warm part of south-west Africa, and is frost-sensitive. The scaly, fibrous, 175 mm (7 in) diameter bulb seems huge for such a small flower stem, which only reaches 600 mm (24 in). The flowers are clustered for 250 mm (6 in) at the top of the stem, like a hyacinth, and are 25–50 mm (1–2 in) long, funnel-shaped, white with a breath of green, and appear late spring–summer. The sword-pointed leaves, which usually follow the flowers, can reach 300 mm (12 in) in height, and can be 75–90 mm (3–3½ in) wide. Plant in large containers as it is vigorous.
(Pseudogaltonia: false galtonia. *Family*: Liliaceae)

Pterostylis (see *Orchids*)

Pulsatilla (see *Anemone Pulsatilla*)

Puschkinia (striped squill) The Middle Eastern striped squill is a beautiful spring-flowering bulb. The forms most widely grown are *P. scilloides* (*P. libanotica*) and its pure white form, *P. s. alba*. The six-petalled, 12.5–25 mm (½–1 in), star-shaped, bluebell flowers, borne in clusters of nine to twelve, on 50–150 mm (2–6 in) stems, are milky blue, with an intense sky blue vein-stripe down

the centre of each petal. The 60–175 mm (2¼–7 in) tall foliage is 5–12 mm (⅕–½ in) wide, fresh, dark green and strap-like.

The bulbs are planted in autumn, 50–75 mm (2–3 in) deep, 100 mm (4 in) apart, in a neutral, free-draining soil, and massed for effect. Puschkinias make useful pot plants when grown in John Innes Compost No 1 or similar. The treatment for puschkinias is as for scillas.
(Puschkinia: Count Apollo Puschkin, plant collector. *Family*: Liliaceae)

Ranunculus The south-east European (Mediterranean) *R. asiaticus* (plate 164) cultivars, whose 37–100 mm (1½–4 in) wide, many-petalled, puffed-up, frilly blooms radiate so colourfully in spring, are easy to grow.

Prepare the soil well before planting, incorporating well-rotted farmyard manure or compost, or a dressing of complete fertiliser, into the soil. Ranunculuses grow easily in most soils, neutral and limy, provided they are not waterlogged. A cold, badly drained soil will produce yellow, stunted foilage. Weeds, particularly annual weeds, will be a problem. These have to be removed carefully, preferably by hand. Wear gloves as many of the weeds have acid sap.

Grow ranunculuses from tuber-claws, which are available in late summer and early autumn. Plant them March–May, preferably late April–May to prevent early growth, which can result in yellow-foliaged, stunted plants, with 37–50 mm (1½–2 in) of soil above the tuber, and 75–150 mm (3–6 in) apart. It is important to plant them with their claws facing down. Irrigation is required during the growing season if no significant rain has fallen. Use top-quality tubers with plenty of plump claws. Ranunculuses can also be grown from seed sown late summer–autumn.

Rose-petalled, single or double flowers may appear on the 300–450 mm (12–18 in) stems in July in temperate areas. However, ranunculuses bloom mostly from late-August or September on. The colours are rose pink, dusty pink, gypsy red, scarlet, crimson, lemon yellow, mimosa yellow, jersey cream and white, in self (one colour), striped, picotee-edged or flaked.

Should you buy seedlings, you may have to take pot-luck with the colours mentioned, unless specified otherwise. The seedlings will make tubers for the following year. If you buy tubers you can specify named colours, particularly if you are designing a colour scheme.

French-flowered, Paeony-flowered, Persian-flowered and Turban-flowered are strains of ranunculus and are available in favourite colour shades or hues.

The flowers will bloom and the foliage will turn yellow. It is time to dig up the plants, tie them by the foliage like bunches of carrots, store them in an airy, dry, cool place, and let the foliage die back. Remove the tubers, place them in shallow boxes filled with dry, fibrous, loamy soil and keep them in a shady place. Plant out the tubers late-autumn, April–May, and keep them moist but not wet. Remember, if ranunculuses are planted out too early they can suffer a check, from which they will find it hard to recover.

There are other species of ranunculus such as *R. acris plenus* (*flore-pleno*) (batchelor's buttons), golden yellow, 13 mm (½ in), ball flowers on 600 mm (24 in) stems. They are easy to grow. *R. lyallii*, the tall, white-flowered, huge-rhizomed, New Zealand alpine buttercup, has large leaves, and is not easy to grow in Australia. *R. amplexicaulis* is a white-flowered buttercup. *R. montanus* has produced many strains, but the cultivar 'Molten Gold' is a fine hybrid, with double-yellow flowers, that seem to melt into gold during early or late sunlight. There are many more raunculus species and hybrids.

(Ranunculus: 'rana', frog; likes damp places. *Family:* Ranunculaceae)

Rechsteineria (cardinal flower) *Rechsteineria cardinalis* (syn. *Sinningia cardinalis*) **(plate 165)** is the species most readily available. It is from the same family as the African Violet, and needs the same growing conditions. It is an indoor, pot-bulbous plant, requiring a daytime temperature of 23.5°C (74.3°F), and a night temperature of 20°C (68°F). The 150–300 mm (6–12 in), heart-shaped, velvety leaves, whorled in twos or threes around the stem, and the brilliant scarlet, 60–75 mm (2½–3 in), tubular flowers with their projecting, hooded, upper lip, make a brave splash of colour.

New plants can be obtained from leaf cuttings, tuber roots or stalk-tip cuttings. Grow them in African Violet compost. *R. verticillata*, the 'umbrella rechsteineria' grows tiers of leaves, like umbrella spokes, and has pink, purple-spotted flowers.

(Rechsteineria: after Pfarrer Rechsteiner). *Family:* Gesneriaceae)

Rhodohypoxis The South African *Rhodohypoxis baurii* is a late spring and early summer-flowering tuberous plant. It produces solitary, six-petalled, two-tiered flowers, three petals in a tier, inter-spaced. The flowers are flat-topped, 'toy-propellor', star-shaped, 25 mm (1 in) wide, in pink, red or white, and are borne on 80–100 mm (3¼–4 in) stems above the foliage. There are many single-stemmed flowers in a cluster. The 6 mm (¼ in) wide, 60–100 mm (2¼–4 in) tall foliage, is hairy and closed around the flower stems.

Rhodohypoxises are mountain plants. They like sunshine, but will not tolerate hot sunshine or wet soils or too hard a frost. Therefore, control their climate by growing the tubers in pots, planted early spring, just beneath soil level, in a rich, acid peat, free-draining compost. Keep them reasonably dry in winter.

(Rhodohypoxis: 'rhodon', rose-like; 'hypoxis', relation to. *Family:* Hypoxidaceae-Amaryllidaceae).

Rigidella Related to *Tigridia pavonia* (Jockeys' Caps), and indeed has been crossed with it to produce a sterile hybrid. The bright red *Rigidella* species are so similar to the *Tigridia* species, apart from their vivid redness, that the description for one could nearly fit the other. Rigidella is subdivided into four species, which are further subdivided as either having erect petals (*R. inusitata, R. orthantha*) or drooping petals (*R. flammea, R. immaculata*). The three, 25 mm (1 in) long, larger outer petals, interspaced with three, smaller inner petals, form a recurved triangle, like cyclamen petals, with a centre-cup. The flower's reproductive organs, situated in the centre, are fused and reach way out from the petals like a radio antenna.

The leaf blades are ribbed and sword-pointed. *R. flammea* has 1–1.5 m (40–60 in) tall foliage, with up to fifteen, short-lived, flaming red, black-

spotted, yellow-cupped, 'afternoon' flowers. *R. immaculata* has brick red 'morning' flowers; *R. orthantha* has brick red outer petals and yellow cup, 'morning' flowers. *R. inusitata* also has 'morning' flowers.

The hybrid *Tigridia pavonia* x *Rigidella orthantha* has bright red outer petals and yellow inner petals, both generally bigger than the species mentioned above, and shows the influence of the vigorous *T. pavonia* parent.

(Rigidella: diminutive of 'rigidus', rigid. *Family:* Iridaceae)

Romulea *R. bulbocodium* produces 50 mm (2 in) long, up-facing, bell-shaped, variable white, pink, mauve, six-petalled, star-ended, flowers, with light mauve inner and yellow-blotched centre. The flowers settle into the 75–125 mm (3–5 in), nearly cylindrical foliage, like lost stars. *R. bulbocodium* flowers open only in bright daylight, in September–October.

R. crocea is similar to the above, but has mimosa yellow flowers. *R. sabulosa* throws satiny, cochineal red, 75 mm (3 in), bell-shaped, up-facing, star-ended flowers, which are veined yellow on the outside and carry a purple blotch in the centre of the trumpet. There are as many as four flowers to a bulb. This 150 mm (6 in) romulea is used in scree, alpine and rock gardens.

Romulea longifolia (*R. rosea*), a prolific South African bulb, has escaped into our pastures and hills. The bell-shaped, yellow-hearted flowers are purplish pink on the top of the inside and purplish green on the outside. The foliage is clustered as thickly as grass and reaches 112–300 mm (4½–12 in) or more. *R. minutiflora* (*R. columnae*), a European romulea, has also escaped. It has small, 12 mm (½ in), violet flowers, with yellow throats, and greenish stripes on the outside petals.

If romuleas are grown, the shiny bulbs should be confined to pots, planted in autumn, and any surplus bulbs should be destroyed. Seek professional advice from your local Botanic Garden, Department of Agriculture, or Home Gardens Advisory Service for information concerning suitable romulea bulbs.

(Romulea: after Romulus, founder of Rome. *Family:* Iridaceae)

Roscoea These lovely, early summer-flowering, Asian ginger-family bulbs are unique in that the hooded, top-petalled, rose pink, deep pink, purple or pale yellow flowers, look like clusters of resting moths clinging to the 100–300 mm (4–12 in) stems. The most important species is *R. cautleyoides* (*R. cautleoides*) from western China, which has hooded top petals, and clusters of purple or pale yellow flowers, borne on 250–300 mm (10–12 in) stems, amid green-grey, 300 mm (12 in) tall foliage. *R. alpina* from Kashmir is a less prolific flowering plant, but the deep pink crinkled flowers, on 200 mm (8 in) stems, will complement any rock garden. *R. humeana*, from China's Yunnan Province, bears masses of rose pink, rose purple 'moth' flowers on 600 m (24 in) stems.

The winter-dormant, fleshy, short-rhizome roots should be planted 150 mm (6 in) deep, 100 mm (4 in) apart, in a lightly shaded, cool climate, and well-drained sandy, peaty or leaf mould, neutral soil. They are not reliably hardy and succeed best in climates that have mild winters and cool, humid summers. All the above roscoeas make good rock garden specimens. Propagation is by division or by seed.

(Roscoea: W. Roscoe, founder Liverpool Botanical Gardens. *Family:* Zingiberaceae)

Sandersonia The summer-flowering, South African *Sandersonia aurantiaca* is called the golden, climbing lily of the valley, and succeeds in partial to mottled shade, or in sunlight in cooler areas.

A tuberous plant, it produces six-petalled, 25 mm (1 in) long, hanging-lantern blooms. However, the petal tips are the barest amount free, and flare outwards. The 'chef's hat' flowers are mouth facing-down, open-throated, orange-yellow, nodding, and borne singly on 25–30 mm (1–1¼ in) delicate stalks, which arise in the leaf-axil joints. The 300–750 mm (12–30 in) stems produce leaves, alternately along their length. The stems are erect, branching and upright (although they may need supporting with canes). The 50–100 mm (2–4 in), stem-clasping leaves are lance-like, parallel-ribbed, narrow, pointed, and send out hooked, clinging tendrils at the tips.

Sandersonia is frost-prone and may have to be protected in some areas. It demands a well-drained, preferably sandy, loam soil. It flowers late spring–early summer. Propagate by seed or by division of the tubers. Plant the finger-like, three-pronged tubers with the prongs pointing down, just below soil surface level, 300 mm (12 in) apart, during their dormant period late autumn–late winter–early spring, depending on the severity of the winter. In frost-prone areas grow the tubers in pots in frost-free conditions, and plant them out in the garden when the risk from frost is over. Bring them back inside after flowering but before frosty conditions occur. Sandersonias make excellent pot plants. Use deep pots, and John Innes Compost No 1 or similar.

(Sandersonia: John Sanderson, Horticultural Society, Natal. *Family:* Liliaceae)

Sanguinaria The North American *Sanguinaria canadensis*, the bulbous bloodroot, a variable member of the poppy family is, apart from being a useful medicinal herb, an attractive, blue, grey-green, heart-shaped or lobed, deciduous, foliaged plant, the 200 mm (8 in) tall by 150–200 mm (6–8 in) wide leaves following the flowers. The plant yields orange-red sap from all its parts, hence its common name. The 25–75 mm (1–3 in) wide, single or double, waxen, white, cup-shaped, stamen-filled, poppy flowers, are borne singly on 150–225 mm (6–9 in) stems in early spring. Each stalk bearing one double flower, has eight to sixteen petals, borne in two to four tiers, and numerous stamens, which are followed by 25 mm (1 in) long seed capsules.

Sanguinaria needs a colder climate and a cool, shady, acid, free-draining soil, rich in organic matter such as peat or leaf mould. Propagate by dividing the cylindrical, finger-thick rhizomes of the double-flowered varieties in autumn–early winter after the root has matured, or by sowing seed of the single flowers. Plant the rhizomes just below soil level 50–75 mm (2–3 in) deep, 225 mm (9 in) apart. One rhizome bud usually only developes one hand-shaped leaf, with three to eight lobed, wavy-edged or toothed 'fingers'.

(Sanguinaria: blood; orange-red sap. *Family:* Papaveraceae)

Satyrium There are over one hundred *Satyrium* species of hooded, ground orchids, originating from the warmer parts of South Africa, India and

China. They have typical, 12.5–25 mm (½–1 in), orchid flowers except for the two spurs, like pink, yellow or orange Chinese moustaches, that hang down under the hooded flowers. The fleshy, flower stems can be 300–850 mm (12–33 in) tall depending on species.

Satyrium princeps, 850 mm (33 in) tall, has carmine pink, hooded flowers, in spring–early summer that cluster around the top 250 mm (10 in) of the stem like a hyacinth cluster. The fleshy leaves, which are pressed close to the ground, are 150 mm (6 in) long, rounded, but tip-pointed, and deeply ribbed; the ribs running up from the base of the leaf to the top. *S. carneum*, 450 mm (18 in) tall, has white blooms with deep pink base, and some fading to pale pink around the outside. It flowers in spring or early summer. *S. coriifolium*, 300–450 mm (12–18 in) tall, has yellow flowers and blooms spasmodically spring–early summer. *S. c. aureum* has yellow-orange, suffusing to crimson, flowers.

Satyriums require ample water while they are growing and a dry soil during their dormant period. This is complicated by the fact that some plants are active in winter and dormant in summer, while others are active in summer and dormant in winter, according to species. Satyrium ground orchids are frost-sensitive, and therefore confined to the heated greenhouse, meaning a 12.7°C (55°F) winter temperature, or a warm climate similar to what they would expect in their native habitat. As their culture can be complicated, it is better that they be grown in pots, in a free-draining, peaty, fibrous, loam compost, with clean, sharp sand added. Generally, the tubers can be divided at the end of the dormant season, just prior to the tubers breaking into active growth.

(Satyrium: Greek name for orchid. *Family*: Orchidaceae)

Sauromatum (lizard flower; voodoo lily) The spring-flowering *Sauromatum venosum* (*S. guttatum*) is a member of the arum or calla lily family, and is found in a line from western South-East Asia to tropical Africa. It is called lizard flower because the barred, speckled-brown throat of the 'flower' spathe is similar to the skin of a lizard. The narrow, 75 mm (3 in) wide, 300–600 mm (12–24 in) long, open-funnel, tapering, 'flower' spathe appears before the leaves, and seems to arise directly from the soil, although it does have a short stem.

The tubers, which can be 150 mm (6 in) in diameter, will flower without soil, living off the stored nutrients in the tuber. This is how it gets its name voodoo lily: a plant that grows and flowers without contact with soil must be bewitched. However, it should be planted immediately after growing this way or it will deteriorate. Also it exudes an unpleasant smell, like rotten meat.

The tall, 300 mm (12 in), true-flowering spadix rises like a pointed bamboo shoot from the centre of the speckled spathe. The minute, female flowers are clustered at the base of the spadix, with the male flowers clustered just above. The 300 mm (24 in) tall, main leaf arises after the flowers, and 'swan necks' under the weight of the leaflets, which are 150 mm (6 in) long and 60 mm (1¼ in) wide, deep green, and arranged as six to twelve leaflets, like a fan, on the curved stem. The leaflets stay green all summer. Tubers should be planted in spring just below ground level in a well-drained soil, 225 mm (9 in) apart. Propagation is by removing young tubers from the parent plant.

It is best grown in pots in hard-frost areas, and protected during winter.

(Sauromatum: 'saurus', lizard. *Family*: Araceae)

Schizostylis *Schizostylis coccinea* is called Kaffir lily, winter ixia, crimson flag or river lily. This South African plant bears up to twelve, glossy scarlet, (or sometimes clear pink), 50–60 mm (2–2½ in), star-shaped, six-petalled flowers, that form like small gladioluses at the top of 450–900 mm (18–36 in) stems. The flowers last for three or four days before being replaced, further up the spike, by others.

Schizostylises, found naturally by the sides of Drakensberg mountain streams, prefer a rich, moist soil, deeply dug, into which well-rotted compost has been added. The site can be in full sun or partial shade. Plant the fleshy, rhizomatous roots, which must not be allowed to dry out while out of the ground, in late winter–early spring, 50–75 mm (2–3 in) below soil level, 150–225 mm (6–9 in) apart. The 300–900 mm (12–36 in), persistent, grass-like, but limpish foliage, will thrust through the soil, and this in itself is attractive. Between late summer and winter, the scarlet, crimson or pink flowers will appear. Schizostylises need to be watered, if no appreciable rain has fallen, during their active growing period.

Kaffir lily cultivars are: *S. c.* 'Mrs Hegarty', a deep, rose pink, floriferous hybrid; *S. c.* 'Viscountess Byng', a free-flowering hybrid, has translucent, shell pink blooms, with deeper pink centres.

Kaffir lilies spread quickly and should be contained. Any rhizomes surplus to your needs should be destroyed. They make excellent pot plants and should be planted in deep pots, in moist, peaty, organic-rich compost. Keep the compost moist while the bulbs are in active growth, and pleasantly damp at other times.

(Schizostylis: 'schizo', cut; 'stylos', style; divided style. *Family*: Iridaceae)

Scilla (squills) European–Asian scillas prefer a cool, semi-shaded position, where they can colonise for many years. They are successful in the cool, temperate areas of the southern states. There are up to ten popular species and hybrids. Scilla flower-bells are not fused together as in the English or Spanish bluebell, but are more open and star-shaped. (see under *Endymion* species). *Scilla natalensis* is a South African species that will tolerate a warmer climate.

The soil should be slightly acid, well-drained, deeply dug, with well-rotted compost added. Do not use fresh manure as the bulbs object strongly to it. Scillas like ample water during their growing season, and if starved of water will have restricted flowering, or no flowers at all.

The depth to plant the European–Asian bulbs depends on their size, which can range from 13–75 mm (½–3 in). The larger, 50–75 mm (2–3 in) diameter, Peruvian (actually Mediterranean) squill bulbs can be planted 88 mm (3½ in) deep, so that the tops are just below ground level and 300–450 mm (6–9 in) apart. The second size squill bulbs can be planted 75 mm (3 in) deep and 100 mm (4 in) apart. The smallest squill bulbs can be planted 50 mm (2 in) deep and 75 mm (3 in) apart. Plant the bulbs early autumn to May. Leave the bulbs to colonise, as they resent disturbance.

Scillas, if planted in the open garden, can have compost or well-rotted manure spread over the

ground where they are growing. Rain or irrigation water will wash the goodness down into the bulb roots.

In late winter–early spring, star-like, leaf rosettes will form at ground level, followed by the dark, thick, glossy green, strap-like leaves. The 75–300 mm (3–12 in), flower stems will appear decorated with swarms of star-shaped flowers. The flowers can be purple, blue, white or light brown depending on the species or hybrid.

S. bifolia, the 75–150 mm (3–6 in), two-leaved squill (although it can have three leaves), originates from southern Europe to Turkey and is the first to appear. The 20 mm (¾ in) flowers are variable from pale violet to deep blue, or reddish blue or white, star-like and flat. The centre stigma and style are prominent and crown the centre. It grows profusely in cold climates, but is less active in the warmer areas.

S. peruviana (**plates 167-8**), from south-west Europe, has a 60–125 mm (2¼–5 in) wide, and 60–250 mm (2¼–10 in) long, bee-swarm head of fifty to a hundred, 13–20 mm (½–¾ in) flowers, which appears from the floppy, coarser foliage on a 180–300 mm (7–12 in) scape (flowering stem) in late spring. The flowers are dark blue, sometimes purple or white, and occasionally pale brown, depending on form or variety. *S. hughii* (**plate 166**) is a larger, blue-flowered form.

S. sibirica, from southern Russia and Turkey, is an early-flowering, variable scilla that does well in the colder climates. The 75–150 mm (3–6 in) flower stems produce three or four, sky blue, lilac or wistaria-violet flowers. It is the hybrids of *S. sibirica* that produce the desirable, purple flowers. *S. s.* 'Spring Beauty' is a favourite hybrid.

S. miczenkoana (*S. tubergeniana*), has 30 mm (1¼ in) pure white or 'breath of blue' flowers, with a dark blue middle stripe on each petal and short, thick, strap-like leaves. *S. autumnalis* produces 150 mm (6 in) flower stems on which are borne minute, reddish purple, pink, black-tipped star-flowers in late summer.

S. natalensis, an early, summer-flowering species, has a tall, 300–900 mm (12–36 in), raceme flower-spike, which contains scores of starry, blue (sometimes white or pink), 14–20 mm (½–¾ in) diameter flowers. Each flower is carried on a slender, 25–50 mm (1–2 in), blue stalk. The spike arises from a dumpy rosette of stiff, wide, pointed leaves. Eventually the foliage will reach a length of 600 mm by 70 mm wide (24 in x 2¾ in). The large, 100–125 mm (4–5 in) diameter bulb, grows with its head, neck and top of 'shoulders' above ground level. Plant the bulb the same way, but make sure it is seated firmly or it will topple. Bulbs can also be planted in a deep, 150 mm (6 in) pot, with the bulb neck and shoulders above the level of the compost.

European–Asian scillas make excellent pot plants, one large bulb to a 125 mm (5 in) pot; two or three of the smaller bulbs to a 125 mm (5 in) pot. Use rich, organic, free-draining compost.

(Scilla: ancient Greek name. *Family*: Liliaceae)

Sinningia (see *Gloxinia*. *S. cardinalis* see *Rechsteineria*)

Sisyrinchium (prairie blue-eyed grass; satin flowers) The spring-flowering, North American sisyrinchiums have cupped, six-petalled, 12–20 mm (½–¾ in) wide, star-shaped blooms that appear towards the top of 200 mm–1 m (8–40 in) stems, amid pointed, grass-like foliage. The flowers vary between species from white to yellow to red to intense blue, and between one and many. Most of the blue and white species and hybrids have a darker blue stripe down the centre of each petal.

S. bermudiana has deep blue flowers, with darker blue veins and yellow throats, borne on branched, flattened, 400 mm (16 in) stems. The leaves are 300–600 mm (12–24 in) long and 6 mm (¼ in) wide. *S. douglasii* is known as the striped purple spring bell; the flowers being reddish purple, 30 mm (1¾ in) wide and only one or two blooms borne on rigid, 150–300 mm (6–12 in) flower stalks. *S. douglasii album* is a white form. *S. campestre*, is the pale blue, blue-eyed prarie grass. *S. iridifolia* (plate 169) has pale yellow, star-shaped flowers, with blue-grey, iris-like foliage. *S. californicum* has yellow, six-petalled flowers, borne above daffodil-like, blue-grey foliage.

Sisyrynchiums require a well-drained soil and a sunny situation. The tuberous rootstock is planted 100–150 mm (4–6 in) deep, late autumn and early spring, 100–225 mm (4–9 in) apart, depending on the size of the tuberous rootstock. Propagate by division of the rootstock or by seed. (Sisyrinchium: an ancient Greek name. *Family:* Iridaceae)

Smithiantha (temple bells) These 37 mm (1½ in) long, red, orange, yellow, pendulous, spotted, bell-shaped, exotic-looking flowers, are aptly named temple bells. However, the opal-flashed, heart-shaped, velvet foliage also takes the eye.

Smithianthas are related to the African Violet, and are greenhouse plants. They require a 16–18°C (60–64°F) night temperature, and a 50–60 per cent humidity. The height of these plants varies with the species, but it will reach 375 mm (15 in).

Cornell University has researched smithianthas and has developed some magnificent hybrids: 'Carmel', 'Abbey', 'Cathedral', 'Santa Clara', 'Vespers', 'San Gabriel', 'Cloister', 'Santa Barbara' and others.

Propagate smithianthas using rhizomes, planting one piece to a 75 mm (3 in) pot, in African Violet compost. You can also take leaf cuttings during the plant's active growth period. Species are *S. cinnabarina*, *S. multiflora* and *S. zebrina* (plate 170).

(Smithiantha: Mildred Smith, English plant collector. *Family:* Gesneriaceae)

Soldanella The flowering stems of these early, spring-flowering, miniature, European, alpine-meadow bulbs grow 50–100 mm (2–4 in) tall, and produce solitary (sometimes two), 8–13 mm (⅓–½ in) long, hanging, purplish pink-fringed-bells. These dainty flowers remind one of naughty children, heads bowed, with their hands behind their backs.

The leathery, glossy foliage is roundish to kidney-shaped, short-stemmed and lies flat on the ground.

Soldanella pusilla, pale lavender, reddish violet, prefers a cool, well-drained site in a rock garden. Plant the bulbs 50 mm (2 in) deep, 50 mm (2 in) apart, or in pots, using slightly acid, gritty, free-draining compost.

(Soldanella: 'soldo', a coin; alluding to leaf form. *Family:* Primulaceae)

Sparaxis (harlequin flower; wand flower) The bright flowers of the South African sparaxis are borne on 250–300 mm (10–12 in), wiry, bent stems. Sparaxises have at least three, sometimes four or five, colours to one flower. The 37–50 mm (1½-2 in) wide, funnel-shape base, open-mouthed, starry, six rounded-petalled blooms, usually have a yellow-star centre-splash, surrounded by a thin, smudgy, ring of black, then white, pink, orange, red or purple petals. However, the centres can vary in colour. On sunny days, in September–October, they throw their petals open wide and display a 'Joseph's coat of many colours'. Ironically, the bright scarlet, self (single) colour, is a much sought after hybrid. On cold or dull days harlequin flowers close up and look insignificant.

Sparaxises have to be contained as they have become pests in some areas.

Harlequin flowers will grow in sun or partial shade in the cool, temperate states, but they do not grow successfully in the north. There are many superb hybrids of *S. tricolor* (plate 171), in yellow, pink, red and so on, and you should plant these rather than the common species. Sparaxis make good cut-flowers if picked when fully open. They are also easily grown pot plants, planted in John Innes Compost No 1 or similar.

The scaly, tunic-coated corms are planted in early autumn in a well-drained soil, 50–100 mm (2–4 in) deep and 75 mm (3 in) apart. They spread by seed and corms. You can plant them in garden beds, lift them each year after the foliage has died down and replant them the following autumn. Corms surplus to requirements should be destroyed.

(Sparaxis: 'sparasso', to tear; spathes are lacerated. *Family:* Iridaceae)

Spathiphyllum A relative of the calla lily or arum family, and a green or white cousin of the Anthurium lily, although more delicate and fragile-looking. The summer-flowering *Spathiphyllum* species produce attractive shiny, green or white flowers (spathes) with white or creamy yellow centre spadixes (true flowers) which turn greener with age.

Spathiphyllum clevelandii (plate 172) is a great favourite with greenhouse or tropical bush-house owners. *S. wallisii* has a white spathe, creamy yellow spadix, and is also popular. *S.* 'McCoy' grows to 1.5 m (60 in), with 200 mm x 450 mm (8½ in x 18 in) creamy white spathes, or white changing to green as they age. Spathiphyllums are killed by frost and need a shady, moist, warm, humid atmosphere, and a winter temperature that does not go below 15.5–21°C (60–70°F). They grow well in a rich, organic soil containing leaf mould, granulated peat, sharp sand and charcoal.

Although grown mainly for their green or white spathes and colourful spadix, spathiphyllums also produce attractive, luxuriant, pointed, paddle-shape, green leaves, which form an attractive backdrop to the flowers. The spathes vary in size, according to species and cultivars, but can be between 50–250 mm (2–10 in) in height to 50–125 mm (2–5 in) in width. They can be grown in pots. Propagate by dividing the fleshy roots in spring.

(Spathiphyllum: 'spathe', broad blade; 'phyllon', leaf; leaf shape. *Family:* Araceae)

Spiloxene (see Hypoxis)

Sprekelia (Jacobean lily; Aztec lily) The Central American sprekelia bears gorgeous, orchid-faced, six-petalled, velvet red flowers, with brght gold, 'dog's-tongue' anthers sticking out. The three bottom petals are nearly joined at their base to form a tube, and this open tube encloses the anthers. The three top petals are long, narrow and curve well back. The two lower top petals are curved in ram's horn fashion. These remarkable bulbs are tough.

S. formosissima (plate 173) is the species grown most frequently, but there are hybrids. The crimson, velvet, six-petalled flowers are 75–150 mm (3–6 in) long, and borne on 300–450 mm (12–18 in) stems. Sprekelias come into flower late spring in the north, and spring–summer in the south. They just keep appearing, and lasting, for some time. The 300 mm (12 in) leaves are deep green and strap-like.

Plant the 50 mm (2 in) wide bulbs, 75–100 mm (3–4 in) deep, 100–125 mm (4–5 in) apart, in a well-drained soil, in late autumn, in a sunny position in the south, and mottled shade in the north. As they spread they will push themselves out of the ground exposing their long necks. Do not disturb the colony until you have to. Keep them well watered during the growing season, but do not waterlog them. They make ideal pot plants, one bulb to a 100 mm (4 in) deep pot, or three to a 150 mm (6 in) deep pot.

(Sprekelia: after J.H. von Sprekelsen; botanist. *Family:* Amaryllidaceae)

Stenomesson *Stenomesson variegatum* from South America, flowers in late winter, spring, and occasionally summer. Of the daffodil family, it is a variable and frost-prone species. The blue-green, strap-like, 25 mm (1 in) wide foliage can reach to 750 mm (30 in). The flowers, which often appear before the leaves, are borne on 425–600 mm (17–24 in) stems, and are carried in a clutch of four to eight, weeping flower clusters. The flower trumpet is 40–75 mm (1¾–3 in) long and 10 mm (½ in) wide, is tubular, nodding, curved and star-shaped at the mouth. The star-mouth is split 15–25 mm (½–1 in) from tip to base. The colour of the flower trumpet can be pale yellow, soft yellow, buttercup yellow with reddish bands, green-yellow, pink with reddish bands, crimson, rose or scarlet. The outside petals can be any of the colours mentioned, but usually display a large green blotch or keel.

S. incarnatum has six or more nodding, 100 mm x 25 mm (4 in x 1 in) tubular bell-shaped flowers. The flowers are variable, but usually orange-red, with a blackish green stripe running down each perianth lobe. *S. incarnatum* has many varieties.

Stenomessons can be grown in deep pots, in free-draining compost. They like a well-drained, open site. Plant the bulbs in autumn or during the dormant season.

(Stenomesson: 'stenos', narrow; 'messon', middle; shape of the flower tube. *Family:* Amaryllidaceae)

Sternbergia (yellow autumn crocus; lily of the field) A lovely Middle Eastern flower, thought to be the original 'lily of the field'.

Sternbergia lutea (plate 174) is a 50 mm (2 in) long, 20 mm (¾ in) wide, mimosa yellow, up-facing, funnel-cupped, crocus-like flower. It arises, with the foliage, usually on 100–150 mm (4–6 in) stems; sometimes 200 mm (8 in). The flowers barely surmount the dark green, channelled, quilled, 7 mm (⅓ in) wide, leaves. The leaves appear when the autumn rains begin;

usually March–April. The flowers will die back when winter approaches, and finally the leaves will shrivel before summer when the plant becomes dormant. *S.l.* s.sp. *sicula* is a narrow-leaved form.

S. colchiflora is a southern European autumn crocus which is not as attractive as *S. lutea* but has similar, smaller, characteristics. *S. clusiana* produces 40 mm (1½ in), mimosa yellow flowers, borne on short stems. The leaves are approximately 300 mm (12 in) long and 10 mm (²⁄₅ in) wide.

Sternbergia fischerana is a spring-flowering species from southern Europe, and has bright, mimosa yellow, 25 mm (1 in) long, 25 mm (1 in) wide, cup-shaped flowers. It is also available as a double-petalled form. The foliage is fine, wiry, with white mid-rib. *S. candida* is the unique, white, spring-flowering sternbergia. The flowers, which are borne on 125–200 mm (5–8 in) stems, are star-like and up 50 mm (2 in) wide.

Plant the shiny black or brown bulbs, after the foliage has died down, usually January–February, 100 mm (4 in) deep and 125 mm (5 in) apart, in a well-drained soil and sunny aspect. They are grown extensively as rock garden plants. The summer sun bakes the ground around the bulbs, which is what they expect in their native environment. Leave sternbergias to colonise. Divide the plants in summer, but only when necessary. Plant them immediately after separating, as they have a short, dormancy period.

Sternbergias make excellent pot plants, planted in deep pots, in John Innes No 1 compost or similar.

(Sternbergia: Count Kaspar von Sternberg, botanist. *Family*: Amaryllidaceae)

Strelitzia (bird of paradise)   The South African bulbous, strelitzia is named after Charlotte Sophia Mecklenburg-Strelitz, wife of King George III. The shape of the stem and flower is like the neck and head of s crane. Bursting from the head of the bird, between late winter and summer, come three pairs of glossy, gloriously bright, orange 'cockatoo' sepals (false petals), which act as the sighting-sign for pollinating birds and insects. The true flowers are deep blue, and point forward like a jet fighter-plane. Bees and birds enter the cockpit in the blue flowers to seek out the nectar. In their natural state strelitzias are pollinated by birds.

The 350–600 mm (14–24 in) leaves of *Strelitzia reginae* (plate 177) are like canoe paddles, deep green, heavily ribbed. These paddles are borne on 600 mm (24 in) stems, making an overall leaf distance of 950 mm–1.2 m (38–48 in). The 150–250 mm (6–10 in) flowers are borne on solid stems 1.2 m–1.5 m (4–5 ft) long. The roots are tuberous and can be divided in early autumn, after flowering, and planted 1.2 m (48 in) apart. Care must be taken when removing the tuber as it is fragile. Therefore dig well down and well away from the plant, taking much of the fibre surrounding the tuber. It is not easy digging out strelitzia tubers, and you may need help.

S. reginae juncea (*S. parvifolia juncea*) has the same brilliant colours, but smaller flowers. The leaves are narrower, rush-like and generally spatulate (spoon-shaped ends). I remember a huge, 5.5 m (18 ft) *S. nicolai* being shifted rather than lose it in a redevelopment. It was expensive and risky to move. However, thirteen years later the plant is alive and thriving. *S. nicolai* looks like a banana tree, and can reach to 6–9 m (20–30 ft).

The 400 mm (16 in) flowers are bird-like, mainly white with some purple, and faint zebra stripes on the head. Plant as a tree, 3 m (10 ft) from its neighbour.

Strelitzias like a deep, well-drained soil, rich in compost, and yearly topdressings of compost or mulch. Tubers planted recently may not flower for a year or two. They grow in most parts except the colder, frost-prone areas. Strelitzias make excellent cut-flowers, lasting for some time in water. They produce pea-sized black seeds with attractive orange parts, the plants can be grown from seed provided the seed has set. Seeds will take years to reach the flowering stage.

Strelitzia reginae (plate 10) makes a good container plant provided you use a container wide and deep enough to accept the extensive root growth, plus the copious water needed to keep the plant growing well. The plants should be grown in a rich, organic, moisture-retaining compost and fed regularly with liquid fertiliser.

(Strelitzia: after Queen Charlotte. *Family*: Musaceae-Strelitziaceae)

Streptanthera   The 'cuprea' in *Streptanthera cuprea* means copper-coloured, and refers to the gleaming, copper-orange of the outside 'halo' of the flowers. *S. c. coccinea* has wide, fan-like, 300 mm (12 in) tall, grassy foliage, which supports 150–300 mm (6–12 in), wiry, flower spikes. The short-tubed, flat, 50 mm (2 in) wide, six-petalled flowers are round-edged and star-like. The centre-spot of the flower is purplish lilac with blackish anthers, and is surrounded by a 'black pen' dash-dotted ring, which in turn is surrounded by a large band of copper-orange.

S. elegans can be white, tinged blush pink outer, with bright purple centre, above which is a velvety black circle marked with yellow spots, or with a yellow throat, cross-banded purple. South African streptantheras are grown from corms planted 100–150 mm (4–6 in) deep, 100 mm (4 in) apart, February–May. They flower in spring. Streptantheras are similar to sparaxises, the harlequin flowers, but are not as exotic a group. They are given the same treatment. They are hardy, but need a well-drained soil and sunny position.

(Streptanthera: 'streptos', twisted; 'anthera', anther. *Family*: Iridaceae)

Streptocarpus   Streptocarpuses (plates 175-6), with their trumpet-shaped flowers, have leaves like primroses and are called 'Cape primroses'. They are principally cool greenhouse pot plants but can be planted out from pots in shaded, frost-free patios or shadehouses, when the weather is warm enough. They are raised from seed sown autumn or early spring, in John Innes Seed Compost or similar, and grown in a 15.5°C (60°F) minimum temperature greenhouse or frame. Streptocarpuses, although perennials, are treated by many gardeners as annuals. The seed is sown each year, early autumn to late winter, for successional flowering; the plants being discarded after flowering. You don't have to discard them, for you can over-winter the plants in a temperature that does not go below 7–10°C (45–50°F). They take five or six months to flower from germination. 'Weismoor' (winter); 'Delta' (summer).

Streptocarpuses can be divided into four convenient groups: (1) Those plants with well-developed leafy stems and stalked leaves 50–75

mm (2–3 in) long, arranged in opposite pairs. (2) Plants without leafy stems, but having solitary or few leaves, which are at least 75 mm (3 in) long, though usually much more. (3) Mouth of the trumpet (corolla) compressed laterally. (4) Mouth of the trumpet (corolla) not compressed laterally. There are many famous species contained among those groups, and more information may be available from books written specifically on such plants.

The fine seed is scattered thinly over the compost and gently, very gently, watered in, as more seed is lost through careless watering than most other seed treatment. You can cover the seed with newspaper and a sheet of glass to speed up, and control, germination. Covering the seeds is not essential, but you will have to check their water requirements.

Constant examination of the seedbox or punnet is necessary to observe when the seeds show the first signs of germination, or any attack of damping-off disease. The newspaper and glass are removed immediately once the first sign of germination is evident. To leave the glass on will draw up and destroy the seedlings. Prick out the seedlings singly into boxes or punnets when they are large enough to handle. Water them in and control the watering from then on. Don't underwater them and don't overwater them. Let the seedlings grow until they have become large enough to prick out into 50 mm (2 in) pots, and then into 75–100 mm (3–4 in) pots, and then into 100–125 mm (4–5 in) pots, and finally 150 mm (6 in) pots. Grow them on for flowering. They should flower spring, summer or autumn, depending on when the seed was sown. Streptocarpuses do not like to be surrounded by too much cold pot soil; hence the many potting stages. The plants should be well watered, kept cool and shaded from the sun.

Streptocarpus clumps can be divided and repotted in winter or early spring. Plants can also be propagated from leaf cuttings.

(Streptocarpus: 'streptos', twisted; 'karpos', fruit. *Family*: Gersneriaceae)

Tecophilaea (Chilean blue crocus)   A beautiful, rare and hard to grow, late summer-flowering, Chilean, bulbous plant that is rapidly becoming extinct in the wild due to over-grazing and collecting. Tecophilaeas have been saved from extinction by collectors and growers who have cultivated them.

T. cyanocrocus is considered similar to a crocus, but it is from an unrelated family. The 50 mm (2 in) wide, flattish, shredded, French-blue trumpets, are borne on 150 mm (6 in) stems. It prefers a cool, mild climate, and rich, well-drained, sandy loam soil. *T. c. leichtlinii*, with its gentian blue outer and white inner trumpet, is a superb variety. Tecophilaeas are frost-prone. The root-covered bulbs can be planted in pots, using a rich, sandy compost.

(Tecophilaea: Tecofila Billiotti, flower painter. *Family*: Tecophileaceae)

Thelymitra (see Orchids)

Tigridia (jockeys' caps, Mexican shell-flower, Mexican tiger lily)   The Central and South American, summer-blooming, short-lived, 'fancy-iris-flowered' tigridias, have the brightness of jockey caps, and the brilliance of tiger lilies, so they are aptly named. The 75–150 mm (3–6 in) flowers, on 300–750 mm (12–30 in) forked stems, come in various shades of white, yellow, pink, red and purple. Eye-catching 'lightning conductor' stamens stick well out from the centre of each flower.

There are three broad outer, 30 mm (1¼ in) wide by 30 mm (1¼ in) long, rounded, tip-pointed petals, that make a 75–150 mm (3–6 in) diameter,. 'three-cornered hat' triangle. These triangular petals form a 37 mm (1½ in) wide cup in the centre. There are three smaller inner, spoon-shaped, pointed, mottled and blotched petals in different colours. You can have golden yellow outside petals with red and yellow centre bowl, or the outside can be pink, with a deep red and white centre bowl. The foliage is like sabre-bladed, pleated, strong-stalked grass. There are a few alternate stem leaves.

Tigridia pavonia, red, spotted yellow-purple, is the principal species, and from it many hybrids have been produced. *T. pringlei* has orange outside petals, spotted, orange-crimson-yellow inner petals, and purplish brown spotted cup, with a centre stamen that reaches out 75 mm (3 in) from the petals.

Tigridias need to be planted in a sunny position, in a well-drained soil enriched with compost. They must be watered well while growing and flowering. The flowers are fleeting, lasting but a day, sometimes more, sometimes less, but are quickly replaced by other buds and flowers.

Plant the fibrous-coated corms June–September, depending on locality, 75–100 mm (3–4 in) deep and 150 mm (6 in) apart, in clumps of six or twelve, for effect. The corms can be lifted once the plants go into their dormant stage and stored in a dry, airy, cool place — usually in clean, dry sand. They can be planted out the following year. If possible, allow them to colonise an area for several years before dividing them.

(Tigridia; 'tigris', tiger; spotted like one. *Family:* Iridaceae)

Tradescantia (spiderwort) The South American *Tradescantia virginiana* (plate 178) is a useful pot plant, mottled shade border plant or shadehouse plant, but only for the more sheltered, temperate areas. The upright stems, arising from strap-like bulb foliage, have narrow, downy, pointed leaves, and reach to 300–700 mm (12–28 in). Many *T. andersonia* hybrids have been produced with deep blue, purple or pink, but rarely white, 25–50 mm (1–2 in) flowers. The flowers have three petals that form a triangular cluster at the end of the stem.

The fleshy, tuberous roots grow well in a neutral soil, or one that is a little acid. Divide the clumps, if needed, after three or four years, late autumn to winter, and plant the roots 300 mm (12 in) apart. Spiderwort refers to its supposed ability to cure spider bites.

(Trardescantia: John Tradescant, Royal gardener. *Family:* Commelinaceae)

Tricyrtis (toad lily) Summer–autumn-flowering toad lilies are cold climate plants native to eastern Asia, and there are ten known species. They prefer partial . shade and are well suited to shadehouse culture, provided they get the correct amount of shade and sun, similar to the liliums that are native to the Himalayas and eastern Asia.

T. stolonifera has 45 mm (1¾ in) wide, six-petalled, starry, pink and red-purple, spotted lily flowers. The base of each of the three outer petals has small swellings like two breasts. The blooms are borne on many-branched stems. The long, 600 mm (24 in), clambering stems, produce 60 mm (2¼ in) long by 30 mm (1¼ in) wide, deep green, heavily ribbed, stem-clasping leaves at regular intervals up the stem. A fascinating feature of the toad lily is the way the centre anthers protrude out from the flat flower, like the spokes of an umbrella with no cover.

T. hirta has white, star-shaped flowers, spotted purple, carried on 300 mm–1 m (12–40 in) stems. *T. formosana* is similar to *T. stolonifera*.

Plant the bulbs-rhizomes 30–50 mm (1¼–2 in) deep, 300 mm (12 in) apart, in late autumn or early spring (usually early spring) in a cool, moist, well-drained, peaty loam, or in free-draining acid peat compost, in deep, well-drained pots. Propagation is by offsets of the rhizomes or by fresh seed sown in a cold-frame. They adapt to deciduous woodland areas or shaded rock gardens, in a well-drained soil. Tricyrtises must never be allowed to dry out. They can be mulched with well-rotted organic material.

(Tricyrtus: 'treis', three; 'kyrtos', covers. *Family:* Liliaceae)

Trillium (wood lily) Most of the *Trillium* species grown in gardens originate from the eastern and northern United States and Canada, although they are found as far afield as Japan and eastern Asia. Some are ill-smelling. They generally grow in deciduous woodlands and thickets, and need a cool, shady site and free-draining, moist soil, rich in leaf mould or granulated peat.

Everything about the trillium is in threes. The leaves are arranged in threes, like large, 50–75 mm (2–3 in), clover leaves, but instead of being rounded at the top, they are sharply pointed. In the centre of the three leaves appears a three-petalled flower on a single stem, and beneath the flower, and between each petal, is a three-leaved green bract. The flowers can be white, green, yellow, blood red, purple-red or blackish red, and bloom late winter–spring. The stems are 125–600 mm (5–24 in) tall.

Trillium grandiflorum has pure white, or maybe with a breath of pink, 50 mm (2 in) flowers, borne singly on 200–450 mm (8–18 in) fine stems. The 80 mm (3 in) long, 40 mm (1¾ in) wide leaves are soft green and smooth-edged. *T. erectum* has red-purple, three-petalled flowers with three, meadow green bracts beneath and between them. The flowers are variable, with green and white forms being collected. The leaves are bright green and smooth-edged. *T. sessile* has blackish red or blood red flowers, and 75 mm (3 in) long, and 75 mm (3 in) wide, deep green, mottled, wavy-edged leaves. *T. x luteum* (*T. s.* var. *luteum*) is a yellow form which has meadow green, mottled leaves.

Wood lilies can be grown in pots in the shadehouse using a leaf mould compost. They are easily propagated from seed but take two or three years to flower. They can also be grown from the warty, sausage-like rhizomes, planted 50–75 mm (2–3 in) deep, 150 mm (6 in) apart, in early autumn.

(Trillium: 'trillix', leaves and flowers in threes. *Family:* Liliaceae)

Triteleia (*T. hyacinthina*; *T. ixioides*; *T. laxa*; *T. lutea* see *Brodiaea. T. uniflorum* see *Ipheion*).

Tritonia South African tritonias have upright-veined, sword-shaped leaves. The two rows of 50 mm (2 in) wide, funnelled, six-petalled, cup-shaped flowers are clustered along 300–600 mm (12–24 in), wiry-reed stems like freesias. The flowers can be obtained in shades and hues of white, yellow, orange or red. There are many lovely hybrids available.

T. crocata, also known as 'Blazing Flame', is a popular tritonia. It has 50 mm (2 in) wide, blazing orange-red, loose-petalled, cup-shaped flowers, with yellow-marked throats and purple anthers, borne on 300–600 mm (12–24 in) stems. These blooms glow like embers in early or late sunlight. Hybrids have been produced with deeper orange flowers. *T. delicata* is a delicate, diffused white. The bright orange, yellow-flecked species, *T. hyalina*, is also grown, but mostly it is a smaller plant with narrower flowers. A batch of seedlings may produce some startling red or orange flowers.

Plant the 25 mm (1 in), flattish tritonia corms during autumn, 12.5 mm (½ in) deep and 150 mm (6 in) apart, in a well-drained soil, in an open site that receives ample sunshine. Tritonias will flower spring–early summer when the orange luminescence of the flowers will smoulder. They make good, reasonably long-lasting, cut-flowers. **(Plate 179)**

Plant tritonias in groups on their own for effect. They can be left to colonise, but contained, or they can be lifted and dried off. Treat as you would gladioluses. Destroy any surplus corms.

(Tritonia: 'triton', weathercock stamens. *Family:* Iridaceae)

(Tritonia *crocosmiiflora*; For information on *Montbretia; Crocosmia* or *Tritonia crocosmiiflora*, see *Crocosmia x crocosmiiflora*.)

Tropaeolum The spring or summer-flowering, Mexican–Peruvian tropaeolums have trumpeted and spurred blooms. They like damp, cool, shady, growing conditions, and neutral to acid soil. Given such treatment, they will produce a flame scarlet or yellow curtain of flowers. They clamber as high as 5 m (17 ft).

The tuberous-perennial *T. speciosum* is grown for its vivid, 37 mm (1½ in) long, orange-scarlet blooms. The flowers are similar to the common nasturtium, long-spurred, 'red witch's hat' flowers found growing like weeds in many gardens. The leaves are five-lobed, like a rounded sheriff's badge.

T. polyphyllum sends out 400–500 mm (16–20 in) trailing or clambering stems. The mimosa yellow, 18 mm (¾ in), trumpet flowers are massed over the plant. *T. tuberosum* has 25 mm (1 in) long, pink-red trumpets with yellow throat, and deep green, white-ribbed, five-lobed leaves. It produces new tubers annually. *T. azureum* is the 32 mm (1¼ in), lilac-blue flowered tropaeolum, and prefers a warmer site than the others mentioned.

Plant the tubers in a well-drained soil after the risk of frost is over, 87.5–150 mm (3½–6 in) deep, well apart to allow for top spread. You can lift the tubers in frost-prone areas and store them like dahlias. Take care as the narrow, branched tubers are usually brittle. They can also be grown from seed.

(Tropaeolum: 'tropaion', trophy. *Family:* Tropaeolaceae)

Tuberose (see *Polianthes*)

Tulbaghia A South African plant named after Ryk Tulbagh, early Dutch Governor of the Cape of Good Hope.

The 400 mm (16 in), reed-like stems of *Tulbaghia violaceae* erupt with a small, 62.5 mm (2½ in), explosion of twelve or more tubular, starry, purple-violet flowers. it makes a good border plant. *T. violaceae*, although long-lasting, has an objectionable garlicky smell when picked, which precludes it from being used as a cut-flower. It does make a good pot plant. The 300 mm x 6 mm (12 in x ¼ in), 'daffodil' foliage is grey-green, although variegated leaf forms are available. *T. v.* 'Silver Lace' is an excellent hybrid. *Tulbaghia fragrans* (**plate 180**) has sweetly scented lilac-mauve blooms, borne in clusters of twenty to thirty at the top of 300–450 mm (12–18 in) stems. Its leaves are larger than *T. violaceae*. There are hybrids available.

Tulbaghias are easily grown from the fleshy, rhizome-like rootstocks, planted in a rich, organic, loamy, well-drained soil, 50–75 mm (2–3 in) deep, 150–225 mm (6–9 in) apart, in autumn. They prefer the warmer temperate states. Increase stock by division of the fleshy rootstocks. (Tulbaghia: after Governor Ryk Tulbagh. *Family:* Liliaceae)

Tulipa (tulip)   The name 'tulip' is derived from the Persian 'dulband' or Turkish 'tulbant', meaning turban. Look at a tulip bulb and it resembles a turban. The tulip's mythological origin is one of supreme sacrifice for a loved one. Farhad, a Persian prince, loved the beautiful maiden, Shirin. A jealous rival told Farhad that his beloved Shirin was dead. Farhad mounted his fastest stallion, rode over a cliff-face and was killed. Out from Farhad's spilled blood sprang red tulips — a symbol of perfect love.

The tulip originated on the line encompassing Turkey, south Russia, north Iran, north Afghanistan, Kashmir and south-west China. It was greatly prized by the Persians and Turks; indeed, the Turks had an official tulip register before Shakespeare was born. There are over one hundred tulip species, with more being found.

It was the Dutch who, in the seventeenth century, developed the tulip from its species origins to the superb hybrids we take for granted today. It is interesting to note that in February 1637, a bulb 'Viceroy', weighing 35 grams (1¼ ounces), was sold for 4200 florins ($2000). Tulips have been split conveniently into groups as outlined below. **(Plates 181-7)**

Early Tulips
Single Earlies These are the earliest of the tulips and are available in all shades and hues except blue. The upturned bells are small, single cups on 300–450 mm (12–18 in) stems. Some are scented. 'Emperor's Crown' is an ancient, red and yellow.
Double Earlies The 100 mm (4 in) flower-cups are as full as a blown rose. They are borne on 300 mm (12 in) stems. They make fascinating edging plants or window box plants, and are available in all shades of white, yellow and red. 'Peach-Blossom' is a mouth-watering pink.

Mid-Season Tulips
Mendels (Duc van Tol x Darwinii) Not sturdy tulips, as even the slightest breeze seems to fluster them. They demand a sheltered spot. The white, pink, red, yellow or apricot cups, carried on 400 mm (16 in) stems, are a delight. 'Apricot Beauty' is eye-catching.
Triumphs A cross between single-earlies and darwins. These are strong-looking blooms, with sturdy, 450–500 mm (18–20 in) stems, and luminescent-edged petals. They are useful plants.

They are not as tall as the darwin tulips, nor do they possess their stately quality, even though they have darwin, plus mendel, blood. 'Garden Party' is a superb, pink, white-edged example.
Darwin hybrids (Darwin x *fosterana*) While not as tall as the darwins, 600 mm (24 in), these hybrids possess other good darwin characteristics, and also the striking red from the mother bulb, *Ta. fosterana*. However, 'Elizabeth Arden' is a lovely, delicate pink.

Late Tulips
Darwins These regal, tall, 600–650 mm (24–26 in) tulips with their glistening cups are available in all shades and sheens except blue. The Darwin is a great parent, siring such winners as the Mendels, Rembrandts, Lily-flowereds and Darwin hybrids. Darwins grow to perfection massed in beds of their own, preferably all one colour: white, ivory, silver-pink, flaked pink, rose, cochineal, salmon, scarlet, currant red, plum, magenta, violet, sulphur yellow, lemon yellow, buttercup and so forth. Bred by E.H. Krelage and named after Charles Darwin. 'Clara Butt' is a pink favourite.
Lily-flowered Lily-flowered tulips have nipped-in waists, with pointed petals, similar to a fairy king's many-pointed crown. These graceful tulips, originating from *T. retroflexa* stock, are 500–550 mm (20–22 in) tall. As they grow older the petals reflex, exposing the centre to the sun.
Single-late Cottage These are similar to single-early tulips, but flower later. They are slender, sometimes pointed, tulips, the sort one would expect to find in the gardens of quaint English thatched cottages. The blooms are variable, being oval, egg-shaped or plum-shaped, and carried on 500–550 mm (20–22 in) stems. The most unusual is the 'green-tulip', *T. viridiflora*, which has pointed, yellow-edged petals, and green from where the stem has filtered into the petals. 'Aristocrat' is red and green. 'Greenland' is rose pink and green. 'Princess Margaret-Rose' is golden yellow, red-edged.

Double Late The petals of this double tulip cluster in a group like a large, decorative dahlia or peony flower, each bloom up to 75–100 mm (3–4 in) across. They are not necessarily a good open-bedding plant, although massed in borders, troughs or window boxes, they look superb. Huddled together in a large bed, they look distressed, with their petals blowing this way and that. Double late tulips are available in white, rose, yellow and orange, either self-coloured, striped or edged. 'Mount Tacoma' is a pure white example.
Rembrandt (broken-bizzare-bybloems). It was a virus disease that sparked off the tulip madness among the Dutch in the seventeenth century. Tulips were popular at that time and expensive, but they were mostly single colours. Then the virus attacked, causing streaking, blotching and barring of the flowers. The price of these infected plants escalated, causing 'tulipomania' or the 'wind trade' (speculators tried to capture the wind). These infected bulbs were moved willingly, all over the then known world. They were white, streaked pink; bronze-orange, streaked bright orange; white, streaked flame; violet, washed white; purple and white flames, on a yellow background; and so on. 'True' Rembrandts range in height from 450–650 mm (18–26 in), with flowers similar to the darwins. 'Cordell Hull' is a lovely pink, named after a former United States Ambassador.

Bizzare is the word to describe broken Dutch–English breeder or broken darwin or broken cottage tulips. The yellow flower ground is marked or striped with maroon, purple, bronze or brown. Bybloem tulips have a white background, with lilac, rose or purple streaks and/or stripes.
Parrot If you can imagine a stately, 500–550 mm (20–22 in) stemmed tulip flower, with its stem bent and petals chewed by the cat, you are close to visualising what a parrot tulip looks like. These tulips, with their twisted, lacerated, split, fringed and curled petals in exotic colours, look more at home in a Sultan's palace than in a backyard. Nobody knows how they originated; all that is known is that they have been cultivated for centuries. Possibly some ancient gardener found them as 'sports' (unusual flowers) growing on a normal plant among his species tulips.

The colours are red, orange-flush; orange and green; black and green; yellow with red stripes; claret violet and brown. Growers have attempted to hybridise parrot tulips but without much success. Parrot tulips have weak necks, which is a fault, but also adds to their unusual charm.
Hybrid Parrot The 'Blue Parrot' tulip is purple and black; the 'Black Parrot', almost midnight; the 'King of the Parrots' is flesh red; the 'Orange Parrot' is luminescent orange; the 'White Parrot' is pure white; the 'Red Parrot' is a gorgeous, flamboyant red. There are many parrot tulips sported from various tulip species and hybrids, namely Darwin, Cottage, Early Single and Breeder. These hybrid parrot tulips are stronger stemmed than the parrot tulips, inheriting this trait from their respective parents.
Dutch Mother Breeder Go to an art gallery and look at the paintings of flowers done by the seventeenth century Dutch masters, to see what these ancient tulips look like. These oval-cupped, 'laquered' tulips lost out to the Darwins in popularity, but with their unusual colour hues and tints, 'breeder' tulips are making a comeback. Breeders can be 900 mm (36 in) tall, and are available in orange, soft orange, old gold, red, purple and coffee, but the base of the tulip is generally stained green, blue-black or black.
English Breeder These blooms look like a small tennis ball, with half to a third of the top cut off. They have a white or yellow base. There are hybrids which have smaller stems and are used extensively in rock garden work and border edging:
Greigii hybrids These vary in height from 200–400 mm (8–16 in) depending on hybrid. They have distinctive purplish blotching and striping on the leaves. The flowers are large and striking. 'Red Riding Hood' has intense red, open blooms. 'Plaisir' has white-yellow open, pointed blooms, with the centre brush-stroked in red. 'Cape Cod' is butter yellow, with a broad, brush-stroke of red down the outside centre of the petals.
Fosterana hybrids They range in height from 350–600 mm (14–24 in). Clear, strong colours. 'Madame Lefeber' is intense scarlet. 'Candela' has butter yellow tight cups.
Kaufmanniana hybrids (water-lily hybrids) These grow 150–200 mm (6–8 in) tall. 'Stresa' has lovely, tight, vivid red and mimosa yellow blooms. 'Heart's Delight' has reddish outer, and white inner cups. 'Johann Strauss' is an open, white-star tulip with mimosa yellow inside the petals.

Tulipa (tulip) species
There are many species growing in the wild, but we shall discuss only a few of the more important:

Tulipa australis A dwarf, 175 mm (7 in) tall tulip from southern Europe. The flowers are butter yellow, with a breath of red on the outside.

Tulipa biflora A 125 mm (5 in) midget that bears up to three, white, star-shaped flowers, with butter yellow centres, on one stem.

Tulipa banuensis A vivid, red-black-throated tulip, which opens right out during sunny weather.

Tulipa clusiana (Lady tulip) An Iranian–Himalyan, delicate-looking bulb, whose 250 mm (10 in) stems produce ivory white, fragile, cups with a pink flush carried on the outside petals and purplish inner blotch. They open into flat, starry blooms. Two 10 mm wide x 250 mm long (²/5–10 in) greeny grey leaves appear at the flower stem base, with other small leaves appearing on the stem.

Tulipa stellata chrysantha It barely climbs above ground level, but the 37.5 mm (1½ in), yellow, red-backed flowers have beautiful golden yellow throats.

Tulipa fosterana The most brilliantly scarlet of all the tulips. It has a black-purple, centre basal blotch touched with yellow. The cultivar 'Madame Lefeber' (Red Emperor) is nearly as brilliant.

Tulipa greigii Scarlet-olive, the most beautiful of the species tulips, with a centre, black basal blotch, margined yellow. Blooms are borne on 160–300 mm (6½–12 in) stems. The foliage is purplish, mottled or striped. The hybrids mentioned previously are magnificent.

Tulipa kaufmanniana This tulip has the same open-flowered, pointed petals, as a water-lily, and that is why it is called the 'water-lily tulip'. The creamy yellow flowers, with a touch of carmine outside, are borne on 325 mm (13 in) stems. There are many hybrids of this species, in various colours such as white, salmon pink, carmine, scarlet and gold-yellow.

Tulipa linifolia This 125 mm (5 in) tall bulb produces grass-like foliage. The loose petals are scarlet, with a bluish to purple-black blotch at the base.

Tulipa polychroma (Biflores section) Has deep, closely folded foliage. The open flower petals, borne on 150 mm (6 in) stems, are white inside and greenish lilac on the outside, with a pat of butter yellow at the centre.

Tulipa praestans A central Asian species that bears orange-red, star-shaped flowers, on 200–300 mm (8–12 in), hairy stems. Parent of the famous cultivar 'Fusilier', *T. praestans* has up to four flowers on one stem.

Tulipa saxatilis (plate 187) The rock tulip. Its pale, lilac-pink flowers, have canary yellow centres. This tulip spreads by stolons, and grows well in parts of South Australia.

Tulipa sylvestris subsp. *sylvestris* (plate 181) This tulip has delicately graceful, 250–450 mm (10–18 in) swan necks that tremble in the slightest breeze. The 50–75 mm (2–3 in) wide blooms are mimosa yellow.

There are many more species of tulips.

Tulips achieve perfection in parts of New South Wales, Tasmania, Victoria and New Zealand. They will only grow in certain areas of South Australia, Queensland and Western Australia. Check with local expert opinion on the suitability of growing tulips in your part of the world, and, most important, the best time to plant them.

Tulips like a rich, well-drained soil, a neutral to moderately limy soil pH. They are planted March–April, some areas May, 150 mm (6 in) deep, to the top of the bulb, in a light soil; 125 mm (5 in) deep, to the top of the bulb, in

heavier soils; 150–200 mm (6–8 in) apart; that is the larger bulbs being the farthest apart. If the bulbs are sending out green shoots, plant them straight away. Use a trowel to plant the bulbs. You can use a dibber (a small section of fork handle that has been sharpened to a point) to plant them in sandy soils. Make sure the bulb is placed firmly against the bottom of the hole, thus ensuring that no air pockets are left under the bulb to collect water. Dress the soil with a complete bulb fertiliser, applied as the manufacturer instructs.

Plant tulips en masse in beds of their own, or over suitable herbaceous plants, the bulbs being selected for colour and height. Tulips of different heights and colours in the same bed do not look good. Label the bulb bed with the bulb's name immediately. For a first-class display, buy top-quality, fresh-stock bulbs from a grower who specialises in tulips. Order them as early as possible, as the best bulbs soon go.

Tulips will flower in spring. Remove the dead blooms and let the foliage manufacture goodness for the bulb. Lift the bulbs once the foliage has died down, and store the bulbs in a cool, light, airy place. Clean up the bulbs, removing and destroying any diseased and soft bulbs. Check constantly for insect and disease attacks while in storage. Separate the largest, young bulbs, as they can be planted next season. The smallest bulbs can be planted in an unimportant section of the garden, and should be allowed to grow, but not flower, for two years. This will help build up the bulb size.

Container-grown tulip bulbs are an excellent way of obtaining quality blooms as the container can be placed exactly where you feel the bulbs will be happiest. Use a suitable free-draining bulb compost and suitable bulb fertiliser.

(Tulipa: dulband; tulbant, turban. *Family:* Liliaceae)

Umbilicus (navelwort) *Umbilicus erectus* is a small, yellow, summer-flowering, creeping-rootstock, crassula plant, native to the Mediterranean coast and Asia Minor. It has an insignificant, but dense, raceme of minute flowers borne on 300 mm (12 in), thin, knuckly stems. Once associated with the *Cotyledon* species, it has (like *Dudleya greenei*, for example) been placed into a different species category.

The 10 mm (²/5 in) long, mimosa yellow, skinny, bell-shaped, ragged-tipped flowers are clustered in groups of twenty-five or more at the top 50 mm (2 in) portion of the flowering stem. The small, flat-funnel, hairy, peltate (the leaf seems balanced on the stem like a plate on a stick) leaves, arise late winter–early spring. These leaves are dimpled in the centre, hence the plant's common name. The thin, knobbly, rhizomatous tubers can be planted 25 mm (1 in) deep, 150 mm (6 in) apart in autumn, in the open in cool areas, or in partial shade, in a free-draining soil. They can also be propagated from seed or leaf cuttings. They can be grown in pots, using John Innes Compost No 1. A collector's plant.

(Umbilicus: navel; depression of the leaves. *Family:* Crassulaceae)

Urceolina (little urn pitcher plant) Can you imagine a 50 mm (2 in) water-urn being tipped upside down but, instead of water, flower anthers come gushing out? This is the little pitcher plant, or in Latin, 'urceolina'. Urceolinas are members of the daffodil family but do not have the typical daffodil flower. They have been cultivated

in the Peruvian Andes for hundreds of years. There are only a few species.

The 40–50 mm (1½–2 in) long, hanging, tubed, pot-bellied, constricted-neck flowers have small, pointed, petal ends, six in number, that flare slightly upwards. These green-yellow-white or scarlet blooms, borne on fine 'u-bend' stalks, erupt from the top of 300–375 mm (12–15 in) drinking-straw stems, like water fountaining out of an upright water-pipe. Urceolinas produce six or more pendulous flowers on the one stem, and are late spring and early summer-flowering, depending on local conditions and species.

U. urceolata has acutely pointed, paddle-shaped, oval, glossy, deeply ribbed, 300 mm (12 in) long, 100 mm (4 in) wide leaves, that remind one of an aspidistra. They appear, usually, with the flowers. *U. urceolata* flowers have a yellow base and green-pointed petal tips, which are edged with white. *U. peruviana* is an intense scarlet with 20–40 mm (¾–1¾ in) long, 10 mm (²/5 in) wide, pot-bellied, constricted-necked, pointed-tipped flowers. It has fewer blooms but its colour makes up for the paucity of flowers. The flower are held on 250–300 mm (10–12 in) strong stems, well above the deep green, 25–40 mm (1–1¾ in) wide, acutely pointed, elliptical leaves which follow the flowers.

Urceolinas can withstand cool, but not hard conditions. In frost-prone areas grow them in pots in a cool greenhouse. The bulbs can be lifted during the autumn dormant stage, before the frosty weather sets in, and replanted early spring, when the risk from frost is over. Plant the brown, 75 mm (3 in), felt-like bulbs in a well-drained soil, in late autumn–winter, with the neck of the bulb 37.5 mm (1½ in) below soil level and 150 mm (6 in) apart. They can be left to colonise an area for some years, or they can be lifted in the late autumn–winter dormant stage and replanted early spring. They can be propagated by using bulb offsets. Urceolinas make good pot plants, particularly if they are tight-potted, in deep pots, in free-draining, bulb compost.

(Urceolina: 'urceolus', small cup. *Family:* Amaryllidaceae)

Urginea *Urginea maritima* is a late summer-autumn-flowering native of south Portugal, the Mediterranean area from southern Spain to Israel, and on to Jordan. These 500 mm–1 m (20–40 in) tall, solid-looking stemmed, pointed, candle-like flower spikes contain a swarm of flowers. The leaves follow the flowers during the winter, remain green until spring, when they die back and the bulb goes into the dormant stage.

On close inspection it is noted that the flower spikes are made up of hundreds of 10–15 mm (²/5–³/5 in), six-petalled, white, star-shaped flowers, with dark brown centre vein and a dark insect-sighting spot on each petal. The leaves are 300 mm–900 mm (12–36 in) long and 100 mm wide. They prefer Mediterranean-like climates.

The bulbs, which are huge, 100–150 mm (4–6 in) in diameter, can preferably be grown in well-crocked, deep pots, but not too large a pot, as they seem to like tight-potting, using John Innes Compost No 1 or similar. They should be controlled, and any surplus bulbs destroyed.

(Urginea: Beni Urgin, Algerian Arab race. *Family:* Liliaceae)

Vallota (Scarborough lily) Early in the nineteenth century, a Dutch ship was battling her way up the north-east coast of England through an aggressive storm. It was a losing battle, and off the town of Scarborough, Yorkshire, the ship gave up the ghost and was dashed on the rocks. The sea, flexing its newly won muscles, attacked the coast and made some inroads, but the coast was too strong and the sea retreated into the deep of the North Sea. The following year, rich red, tempered orange-scarlet flowers, some with white centre spots, were found growing above the shoreline. They were vallota lily bulbs washed up from the shipwreck. The vallota is now known as the 'Scarborough lily'. It sounds romantic, but the truth is that the local Scarborough residents took the bulbs and grew them in their gardens.

Vallota speciosa · (*V. purpurea*) **(plate 188)** is a native of South Africa but will grow well in most temperate areas of Australia, except certain parts of Western Australia and Queensland. The foliage can be deciduous or more or less evergreen in the colder, southern climates, and fully evergreen in warmer climates. In the wild it is known to be deciduous.

Vallotas like a well-drained soil, rich in fibrous compost, and ample water when they are growing well. They are suited to well-drained rock gardens. From among the fan of six to fifteen, 15–25 mm (⅗–1 in) wide, 250–500 mm (10–20 in) long, elegant, strap-like, dark green leaves, a thickish, 300–450 mm (12–18 in) stem appears. At the top of the stem, January to March, blooms appear in clusters of four to six. The flower trumpet is split deeply into six, rich red, tempered orange-scarlet petals. The flowers are 37–50 mm (1½–2 in) long, 75–125 mm (3–5 in) wide at the mouth, with delicate golden anthers and stamens.

V. s. alba is a white form. *V. s. exima* has a white throat with crimson feathering. *V. s. magnifica* has 100 mm (4 in) wide-mouthed, white-edged blooms. *V. s. delicata* is a delicate pink form.

Dead flowers should be removed immediately to avoid their using bulb energy to form seed; that is unless you want the seed for propagation. Many bulbils are produced on the mother bulb, which can be removed and grown on. However, it is best to leave the bulbs undisturbed for some years. Plant the bulbs in winter–early spring, (although some recommend immediately after flowering when the plant is in active growth, or just before flowering), in an open position that gets ample sunlight, 50 mm (2 in) deep and 150 mm (6 in) apart, so that the shoulder of the bulb is just below soil level. Vallotas can be increased by sowing the black-winged seed in pots or seedtrays.

Vallotas make superb pot plants. They should be tight-potted, which means they are placed in sandy compost, in small pots, the roots being allowed to crowd the pot. They are kept well watered, and fed with liquid fertiliser when they are growing well.

Vallota x cyrtanthus An intergeneric hybrid between *Vallota speciosa* and *Cyrtanthus sanguineus*. Interesting plant for the collector.

Veltheimia The late spring–summer-flowering South African *Veltheimia bracteata* (syn. *V. viridifolia*; *V. undulata*) **(plate 189)** produces small, drooping-tube flowers of soft, delicate, 'raspberry juice' purplish pink, with a faint yellow, green ring at the end of each tube. The flower's stamens protrude from the tube, like minute bell-clappers. Veltheimias look like smaller, pink versions of red-hot pokers, or larger, pink versions of lachenalias. The 25–50 mm long, bladder-seed-

fruits are unusual in that they form wide, paper-tissue wings.

The huge bulb will produce a matt-rosette of 75 mm (3 in) wide, wavy, intense, deep green leaves. From out of the leaves a thick, fleshy, mottled, leafless, 450 mm (18 in) stem will appear, at the top of which, like a tubular ice-lolly, will be borne a mass of thirty, forty, even sixty, tubular, 30–40 mm (1¼–1½ in) long, slightly drooping flowers. Veltheimias make superb pot plants — three 50 mm (2 in) bulbs to a 150 mm (6 in) pot — which can be fed with a slow-release granular fertiliser. They also make good cut-flowers.

Veltheimias will grow outside in frost-free temperate places, but usually they are protected from fierce winds, heat and the like, by trees, shrubs or fencing. They prefer light shade. They will not grow outside in the frost-prone, colder areas, but are easily grown in greenhouses.

Veltheimias require a well-drained soil, rich in well-rotted organic matter such as compost or farmyard manure in a site that receives adequate sunlight, and light shade. The 50–75 mm (2–3 in) bulbs should be planted in late autumn, 125 mm (5 in) deep. Be sure, once growth begins, that they receive adequate water. Stop watering when the flowers have finished. The bulbs should be allowed to ripen naturally after summer flowering, as this process is important in maturing them for the following year. They usually receive ample rain during the winter.

Leave veltheimias for some considerable time to colonise an area, as they are snails when it comes to spreading. Hand-weed around veltheimias, taking great care not to damage the bulbs, as they are easily injured. The bulbs can be lifted after the summer flowering and stored in a frost-free area, in pots of compost, and planted out when the risk from frost is over.

There is a narrower, blue-green foliaged species, *V. capensis* (*glauca*), which has 25–35 mm (1–1½ in) long, red-spotted or yellowish flowers, but it is not much grown in Australia.

(Veltheimia: August, Graf von Veltheim, botany patron. *Family:* Liliaceae)

Wachendorfia The South African, early spring–summer-flowering *Wachendorfia thyrsiflora* produces a 1.2-1.5 m (48–60 in) tall, dense, tight, spike of mimosa or buttercup yellow flowers, usually with darker markings on the rear petals. They are 30 mm (1¼ in) wide at the mouth, six-petalled and star-shaped. The 600–900 mm (24–36 in) tall, sword-pointed, mid-green, nearly evergreen, smooth foliage is ribbed and pleated, and forms an attractive 'accent' plant, particularly when in flower. *W. paniculata* is 200–300 mm (8–12 in) tall with apricot yellow flowers, 25 mm (1 in) wide, with darker markings on the rear petals. The leaves are 450 mm x 25 mm (18 in x 1 in) tall, and are slightly hairy. There are up to twenty-five species in cultivation.

Wachendorfias are used as waterside plants. They prefer a rich, organic moist soil and partial shade during summer. They are frost-sensitive and will not survive without proper protection during cold weather. The tuberous roots are crimson red, inside and out, and are planted in autumn 50–75 mm (2–3 in) below soil surface, 225–300 mm (9–12 in) apart, and left to colonise. They can be dug up in autumn and separated.

(Wachendorfia: E.J. Wachendorff, Dutch botanist. *Family:* Haemodoraceae)

Watsonia Watsonias are beautiful scented plants, particularly the newer hybrids **(plate 190)**. Watsonias produce sword-shaped leaves similar to the gladiolus. They have the same flowering-spike habit, although the flowers are different. The 63 mm (2½ in), funnel-base, six-petalled blooms are flatter and more starry, with pointed, brown, glossy anthers in the centre. The petal tips, usually flare outwards. The ribs of the flowers are a deeper colour than the petals.

The corms are planted in a well-drained soil, rich in well-rotted compost or acid peat. Plant only firm, healthy corms 87.5 mm (3½ in) deep and 175 mm (7 in) apart. They will tolerate a wide range of soils and weather conditions. However, they like ample moisture during their growing period.

Leave watsonias to colonise the area until they become overcrowded, then divide and replant the mixture of corms and roots during their dormant period in late autumn, March–May. Destroy any surplus bulbs, as this South African bulb has escaped into the hills and mountains in certain parts of Australia and become a pest.

The newly divided corms will send up their sword blades soon after planting, and by July will have produced a formidable array of foliage. The leaves will reach to a height of 600–900 mm (24–36 in), but they will be surrounded by the flower spikes which will reach to a height of 1.2–1.8 m (4–6 ft), depending on species or cultivar. Watsonia flowers appear in October, although earlier in good years and in the correct climatic conditions. They have a long flowering season. Some evergreen species flower in mid-summer.

There are some fine cultivars available. Many are named after Australian capital cities: 'Adelaide', orange-scarlet; 'Canberra', rosy mauve; 'Hobart', white; 'Melbourne', salmon pink; 'Perth', magenta; 'Sydney', magenta. There are many other hybrids.

W. beatricis is an evergreen species with sun-kissed, apricot red, 50–75 mm (2–3 in) wide, waxy, blooms. It flowers from mid-summer on. *Watsonia marginata*, deciduous, is a fine rose pink species producing a mass of pink, star-ended trumpets, with deeper pink mid-ribs. *W. aletroides*, deciduous, has unusual, sharply bent, tubular flowers, that have a white-pinkish-scarlet glow.

(Watsonia: Sir William Watson, naturalist. *Family* Iridaceae)

Worsleya (*W. procera* see *Hippeastrum procerum*)

Zantedeschia (Arum lily; calla lily) Of the eight South African *Zantedeschia* species, mostly *Z. aethiopica* **(plate 12)**, *Z. rehmannii*, *Z. elliotiana* **(plate 191)** and *Z. pentlandii* are grown. *Z. aethiopica* grows naturally in wet, marshy places and river banks, in areas such as the Drakensberg Mountains. It is variable in its hardiness.

Zantedeschias flower September–December onwards, depending on species and hybrid. They grow easily in light shade in most urban areas, but are vunerable in frost-prone sites, although *Z. aethiopica* foliage, when cut down by frost, may recover. In southern England *Z. aethiopica* tuber rhizomes are planted below soil water level in damp places, to place them below frost risk.

Z. aethiopica, the white arum lily, is widely grown and flowers appear from September to December. Arum lilies are used in floral decoration. The large, pure white, 100–200 mm

(4–8 in) wide, split to the base, overlapping funnel-cup flower (spathe), with its characteristic tapered, recurved, piggy-tail tip at the summit, has a pencil-like golden spadix (true flower spike) in the centre. The spadix is clustered with swarms of minute, male and female, yellow flowers. It grows up through the centre of the funnel, and can protude above the cup.

The white arum lily, or calla lily (calla meaning beautiful), is in great demand as a cut-flower for churches, as its intense whiteness denotes purity. The glossy, deep green, arrowhead-shaped leaves can grow up to 1.2 m (4 ft) tall, but generally they are shorter. Z. a. 'Crowborough' is a popular hybrid. There are smaller forms of Z. aethiopica available, which look fine in a small section of the garden. Z. a. minor 'Little Gem' is 450 mm (18 in) tall.

Z. elliotiana is the yellow arum lily, with transparent-spotted, pleasing green leaves. It is deservedly popular. The yellow, overlapped, piggy-tail, funnel-shaped flowers are 100 mm (4 in) across. The stems reach to 600 mm (24 in). The leaves are paddle-shaped, with characteristic papery blotches on the surface. It requires a moist, but free-draining soil, rich in compost or peat.

Z. rehmannii is the light pink to violet-red arum lily, but it can be variable, having white or white greenish throwbacks. It is, at 275 mm (11 in), smaller than most others and has paddle-shaped, pointed leaves. This height should be remembered when planting Z. rehmannii. Plant in a moist, free-draining soil.

Z. pentlandii (Z. macrocarpa) is a yellow form, but produces a characteristic black-purple spot at its throat. It is considered by some experts to be a form of Z. elliotiana. It has large, green, unspotted leaves and needs good drainage.

Z. albo-maculata is the white, spotted, arum lily. Its name, albo-maculata, refers to the papery spots on its glowing, triangular arrowhead leaves. It has whitish or pale greenish yellow, funnel-flowers with a blotched crimson base, depending on the sub-species. Like all calla lilies it is extremely attractive. Z. a-m. 'Helen O'Connor' has gloriously sun-kissed, apricot peach flushes to its flower spathe. The above like a free-draining soil, rich in organic material.

Growers have hybridised zantedeschias, particularly Z. elliotiana and Z. rehmannii. They have produced magnificent cultivars of white with brownish throats, yellow with brownish red throats, deep red, deep pink and so on.

Plant the rhizome tubers autumn–winter, 100–150 mm (4–6 in) deep and 150–300 mm (6–12 in) apart, depending on rhizome tuber size, in a soil enriched with well-rotted farmyard manure. The soil should be moisture-holding, not permanently waterlogged, but free-draining.

Excessive waterlogging can lead to a disease of the rhizomes, like rotting potatoes, known as Bacterium ariodiae (aroid wet, soft-rot). Once rotting is discovered, scrape it away down to clean fresh tissue and treat the area with a suitable fungicide. However, Z. aethiopica naturalises exceedingly well in many of the wet banks and streams among our shady hills. All I can suggest is that the water-table lowers considerably during summer, allowing the fleshy rootstock to enjoy a long drying period, as opposed to being permanently waterlogged.

Zantedeschias need copious water during their growing season. The leaves will turn yellow in hot sunlight, so select the right planting spot carefully.

Dig up and separate the rhizome tubers when the flowers begin to deteriorate, and replant the largest and healthiest sections, making sure that there are growing buds on each section. Destroy surplus rhizome tubers, as Z. aethiopica is considered a pest in Western Australia.

Zantedeschias can be grown easily from seed sown in autumn for white forms, and spring for coloured forms. They flower usually in the third season.

(Zantedeschia: Giovanni Zantedeschi, Italian botanist. Family: Araceae)

Zephyranthes (Argyropsis) (zephyr lily; storm flower) The up-facing, summer–autumn, occasionally spring-flowering blossoms of the South American zephyr lily, borne singly on each stem, have a trumpet-shaped base, with the mouth of the trumpet split to form six oval-pointed petals. The flowers can appear suddenly after rain, hence the name storm flower. Zephyranthes are mostly deciduous; some are evergreen, becoming semi-dormant in winter.

Z. grandiflora, a native of Central America, bears 75 mm (3 in) long, 75–100 mm (3–4 in) wide, shell pink flowers, with golden, centre anthers, carried on 225 mm (9 in) stems. The large flowers, which can be up-facing or slightly tilted, seem to nestle among the taller leaves. It flowers from summer on.

Z. candida (Argyropsis candida) (plate 192), commonly known as the Australian crocus, is neither a crocus nor Australian, and occasional blooms do appear in summer. However, this Argentinian plant does resemble a white crocus. It is a prolific plant with viable seeds and bulbs, and spreads rapidly. The 50 mm (2 in) long, 37–50 mm (1½–2 in) wide, up-facing, white flowers, borne on 150 mm (6 in) stems, are glossy in sunlight and will, during early morning or late evening, cast rose pink shadows inside their flower funnel-cups. The 400 mm (16 in) leaves, which are longer than the flower stems, are evergreen, wiry, rush-like and persistent. It prefers a rich, moist, but well-drained soil.

Z. citrina, the sweet-smelling, zephyr lily from South America–Caribbean area, throws intense yellow, 37–50 mm (1½–2 in) funnel-shaped flowers, usually one to a 200 mm (8 in) 'drinking-straw' stem.

Hybrids from the above species have been produced, bearing features from both parents, but are, mainly, pale yellow forms. The deciduous, pink, late spring–summer-flowering, Z. grandiflora 'Rosea', is grown in the north, more so than the others mentioned.

Plant the 20 mm (¾ in) spherical bulbs 50–100 mm (2–4 in) deep, depending on the soil. The lighter the soil the deeper you plant the bulbs. They should be 75 mm (3 in) apart and planted during the autumn-winter dormancy. Choose a site that receives full sun, except in Queensland, where full sun is too much and partial shade is required.

Zephyranthes prefer a rich, light, well-drained soil, preferably with well-rotted farmyard manure or well-rotted compost, incorporated. Ensure that the bulbs get sufficient water during the growing and flowering season, but allow them to ripen off late summer. THey can be used as border or rock garden plants.

They make fine pot plants planted in deep pots of John Innes Compost, kept watered well during the summer, and fed with a slow-acting, granular fertiliser.

(Zephryanthes: 'zephyros', west wind; 'anthos', flower. Family: Amaryllidaceae)

Zephyranthes drummondii (see Cooperi pedunculata)

Zigadenus Zigadenus fremontii, is an early summer-flowering plant native to the west coastal, well-drained scrublands of the United States. This lily plant produces a cluster of between ten and fifteen, 13 mm (½ in), pale, creamy yellow, six-petalled, star-shaped flowers on one 180–300 mm (7–12 in) stem. The yellow flower stamens protrude for 5 mm (¼ in) beyond the petals. The variable 120–600 mm (5–24 in), tall foliage, which looks like pale green grass, encloses the flower stem and inflorescence. Z. micranthus is a smaller plant. They are dull and unexciting bulbs, and the thick, rhizomatous bulbous plant is reputedly poisonous to stock and humans.

(Zigadenus: 'zygos', yoke; 'aden', gland; paired petal-glands. Family: Liliaceae)

Zingiber (ginger) Most of us know of the sweetmeat ginger, which is the end-product of the dead rhizome of the plant called Z. officinale. The root is treated in a special way and sold as the expensive sweet. Z. officinale is also a flowering bulb and produces lovely, tight, clusters of yellowish green, open, 'pine-cone' bracts, with dark purple, yellow-spotted lips enclosing the insignificant flowers. The plant is 900 mm–1.2 m (36–50 in) tall, with 150 mm (6 in) long, 17 mm (¾ in) wide, grass-like foliage, poking out at alternate intervals up the stem. Z. spectabile is a more exciting species, having a 300 mm (12 in) flowering stem arising directly from the rhizome, which is an open, pine-cone cluster of yellow-scarlet bracts enclosing what at first sight look like red and black blackberries. The 180 mm (7 in) leaves form alternately either side of the 1.2–1.8 m (48–72 in) foliage stems. The cultivation of zingiber-gingers is as for hedychiums.

(Zingiber: 'zingerberi', ancient Greek for ginger. Family: Zingiberaceae)

Glossary

ACID A soil whose pH is below pH7. Opposite to alkaline.

ACTIVE INGREDIENT Principal ingredient in insecticides, fungicides, etc.

ACUMINATE Long and pointed.

ACUTE Pointed.

AERATION Soil broken up to allow air penetration.

AERIAL BULB Bulb formed in the axil of a leaf and stem.

ALKALINE A soil whose pH is above pH7. Usually limy.

ANNUAL Plant germinates, flowers and dies in one year.

ANNULATE Base splitting into rings (crocus corm).

ANTHER Flower stamen part which bears the pollen.

APPENDAGE Apparently useless attachment to an organ.

AXIL The angle between leaf and stem junction.

AXILLARY Growing from leaf and stem axil.

BIENNIAL Flower grows vegetatively one year, flowers and dies the next.

BASAL Usually applied to a leaf arising from the base of a stem.

BASAL (PLATE) Flat base of a bulb, from which roots grow.

BIFID Something divided into two.

BRACT Modified leaf, usually green and found under a flower or at the flower stalk base.

BULB Underground storage stem (organ), with scale or scales enclosing it, and containing the plant embryo growing point.

BULBIL Usually small, secondary offset bulbs produced, depending on species, on various parts of the plant, which can be propagated.

BULBLET Small or secondary bulbs which form around the mother bulb.

CALCAREOUS Plants growing in limy or chalky soils.

CALYX Whorl of modified leaves (sepals) enclosing the flower bud.

CAMPANULATE Flower shaped like a bell.

CAPSULE A dry seed case (pod), which splits and sheds seed.

CHELATE Holds plant foods (elements) in a claw-like grip, and makes them available solubly in limy soil. Iron chelates; magnesium chelates; Manganese chelates.

CHEQUERED Squarish, more or less regular mottled veining on flowers and leaves.

CILIATE Hair-like fringe.

CLONE A plant produced vegetatively, bearing all the identical genes of the original plant.

COLONY A group of the same plant species.

COMPOST Decomposed organic material.

COMPOST (JOHN INNES) Seeding, cutting and potting soil mixes.

CONTACT HERBICIDE Kills plant through leaf or stem contact.

CONTACT INSECTICIDE Kills insects which absorb through the body.

CORM Solid, not scaly, bulb-like, underground stem (e.g. gladiolus).

COROLLA General name given to the petals of a flower.

CORONA Usually saucer-shaped or a cup-like trumpet inside the outer ring of petals (e.g. daffodil trumpet).

CREST 'Cockscomb' ridge formed on the 'falls' of certain irises.

CROCKS (CROCKING) Usually broken pieces of clay flower pots, placed at the bottom of flower pots to facilitate drainage from the pot.

CULTIVAR A cultivated variety, usually given a non-Latin name, written in non-italic type and enclosed thus: 'Australia'.

CUP For example, a trumpet of a daffodil.

CUTTING A section of stem, leaf or root that contains the necessary plant genes, which can be 'struck' to produce a new plant.

DAMPING-OFF Fungus disease which causes sudden collapse of seedlings.

DILATED Swollen and expanded.

DISTICHOUS Arranged together in two rows; flattened.

ENTIRE A flower petal or leaf with no teeth, lobes or divisions.

FALCATE Shaped like a sickle.

FALL Outer flower petal of an iris, usually falling outwards and downwards.

FAMILY Group of plants containing one or many genus of plants (e.g. Amaryllidaceae contains *Narcissus; Nerine; Amaryllis* etc.).

FILIFORM Thread-like.

FRUIT Seed-bearing part of a plant.

FUNGICIDE Powder or liquid compounded to destroy fungus.

GENUS A group of plants (species), which have common structural characteristics, but are different from other plants. (e.g. *Narcissus*, genus; *Narcissus concolor*, species; *Narcissus bulbocodium*, species; *Narcissus cylcamineus, species). There may only be one known species in a particular genus.

GERMINATION The sprouting of shoots or roots from seed.

HUMUS Well-decayed vegetable mould.

HYBRID A plant cross-bred from two plants of different species or varieties.

INFLORESCENCE Usually refers to flower stem and flowers.

INSECTICIDE Material compounded to control insects.

LANCEOLATE Lance-like, tapered; usually wider towards the middle.

LAX Loosely spaced out.

LEADER Principal shoot of a plant.

LIMY Alkaline soil etc. containing chalk or limestone.

LINEAR Narrow and usually parallel (of leaves).

LOBE Ear-like projection of leaf or flower, etc.

LOCAL (LOCALISED) Plant family or group not widespread.

MOUTH Open end of flower, e.g. daffodil trumpet.

MULCH Topdressing or suitable material used principally to conserve moisture or to supply nutrients to the plant.

NATURALISED Foreign plant establishing itself locally.

NECTARY Nectar-secreting part, usually situated at the inside base of the flower.

NODE The 'knuckle' on the stem from which the leaf grows.

OBLONG Longer than broad, near enough parallel.

OBOVATE Opposite egg-shape. Top broader.

OFFSET Small bulbs produced on the mother bulb.

OVARY The female part of the plant, which following fertilisation bears the seed. It can be 'superior', i.e. borne above the petals, or 'inferior' i.e. borne below the petals.

OVATE Shaped like an egg.

PALMATE Leaves arranged like the fingers of a hand.

PANICLE Flowers arranged in a compound pyramid shape.

PEDICEL Flower stalk.

PERIANTH The 'saucer', showy, outer ring of petals. e.g. daffodil.

PETAL The usually conspicuous, coloured part of a flower. Also known as 'tepals' on certain bulbs.

PETIOLE Leaf stalk.

pH The power of hydrogen in the soil. A convenient symbol to define the acid, neutral or alkaline nature of a soil. pH7 is neutral (pure water). pH7 down to pH0 is becoming progressively more acid. pH7 up to pH14 is becoming progressively more alkaline. (pH = the negative logarithm of the hydrogen-ion concentration in the soil.)

POT-BOUND Roots crammed or overcrowded in a pot.

RACEME A simple arrangement of flowers on a stem, on short, mostly uniform-length, stalks, e.g. a bluebell.

RECURVED Petals curved or flaring 'turk's cap' outwards or backwards.

RHIZOME Usually underground swollen rootstock stem, containing the necessary characteristic bud nodes to produce shoots and roots.

ROSETTE Usually a flat, circular, tightly packed, ground-hugging cluster of leaves.

SCALE Fleshy, overlapping parts of a bulb (e.g. onion).

SELF Flower all one colour i.e. all blue or all red etc.

SEPAL The green 'petals' (calyx) that enclose a flower bud.

SERRATE Leaf or petal having toothed edges.

SESSILE Flower or leaf having no stalk.

SHARP SAND Coarse-edged. Used much in composts.

SPADIX The fleshy, finger-like spike of packed, minute flowers found emerging from the centre of the spathe of calla and arum lilies.

SPATHE Modified papery, funnel-shaped, leafy bract, usually found enclosing the flower spike of calla or arum lilies. Often mistaken for the flower.

SPECIES A group of plants having characteristics in common, but distinct from other groups or species of the same genus.

SPIKE The flowers have no visible stalks, i.e. they are borne directly attached to the stem, such as grape hyacinth flowers.

SPORT Part of a plant distinctly different from the parent; e.g. a viable, red flower on a yellow 'King Alfred' daffodil.

STAMEN Male flower part, consisting of pollen-producing filament-anthers.

STANDARD The usually erect, inner petals of an iris flower.

STIGMA The usually sticky, female part of the flower, which receives the pollen and transfers it down the style to the ovary.

STOLON Underground stem, which produces bulbs from nodes along its length; sometimes clambering above ground in certain plants.

STYLE Connects the female stigma to the ovary.

SUBSPECIES A group of plants within a species, having slightly different characteristics, but not enough to be classed as a separate species.

SYNONYM Previously used name for a plant.

SYSTEMIC Fungicides or insecticides (usually for sap-sucking insects), that are absorbed through the plant's leaves or tissue and translocated throughout the plant.

TENDRIL Usually a leaf extension capable of twining around objects as a means of support.

TEPALS The petals of many of the lily family are known as tepals.

TILTH Soil broken down with cultivator or fork into a coarse, medium or fine structure.

THROAT The upper part of a flower tube.

TOOTHED Leaf or flower edge with teeth-like projections.

TRUMPET The corona of say a daffodil, which is longer than saucer or cup-shaped.

TUBER Swollen underground stem, containing all flower parts; e.g. potato and dahlia.

TUNIC Bulb coating, often netted.

UMBEL Flat-topped, umbrella-spoked flower cluster; flower stalks emanating from a single point.

VARIETY A natural variation of plants in the wild, differing markedly from the species involved.

WATERLOGGING Soil so saturated with water that all oxygen is driven out.

WEEDKILLER Liquid or powder to selectively or totally kill weeds, depending on the weedkiller's composition.

WHORL A 'propellor' group of leaves or flowers, at one level, encircling the stem.

Bibliography

AUSTRALIAN LILIUM SOCIETY. *Liliums—Cultural Notes.*

BETTER HOMES AND GARDENS. Gardening Section, (Editor: S. Macoboy).

BRUNNING'S AUSTRALIAN GARDENER. Angus & Robertson, Sydney, 1983.

CREMLYN, R. *Pesticides — Preparation and Mode of Action,* John Wiley, Chichester, 1978.

DEPARTMENT OF AGRICULTURE. (All States) *Journals and Fact Sheets*

DIX J. F. Ch. *Bulb Growing for Everyone,* Blandford Press, London, 1957.

DOERFLINGER F. *The Bulb Book,* David & Charles, London, 1973.

EVERETT, T. H. *The New York Botanical Garden Illustrated Encyclopedia of Horticulture,* Garland, New York, 1981.

EDWARDS R.G. *The Australian Gardening Book,* Angus & Robertson, Sydney, 1958.

ELIOVSON S. *Bulbs for the Gardener,* Reed Wellington, 1968.

HANCOCK J.N. & CO. (Menzies Creek, Victoria), *Bulb Catalogue,* 1985.

HARRISON, R.E. *Handbook of Bulbs and Perennials,* Printed by Keeling & Mundy Ltd, Palmerston North (NZ), 1953.

HOCKINGS F.D. *Friends and Foes of Australian Gardens,* Reed, in association with the Society For Growing Australian Plants, Sydney, 1980.

LOTHIAN N. *Complete Australian Gardener,* Rigby, Adelaide, 1976.

MORLEY B.D. & EVERARD B. *Wild Flowers of the World,* Octopus, London, 1970.

MOIGNARD B. & GEDYE L.R. *Water Gardens Made Easy,* Southdown Press Melbourne.

MATTHEWS B. *The Larger Bulbs,* Batsford, in association with the Royal Horticultural Society, London, 1978.

OAKMAN H. *Gardening in Queensland,* Jacaranda, Brisbane, 1958.

PESCOTT R.T.M. *Bulbs for Australian Gardens.* Nelson, Melbourne, 1968.

RIX, M. & PHILLIPS R. *The Bulb Book,* Pan, London, 1981.

ROYAL HORTRICULTURAL SOCIETY. *(various journals).*

ROYAL HORTICULTURAL SOCIETY. *Dictionary of Gardening* (4 volumes; plus supplements) (Ed. F.J. Chittenden), 1956.

SEALE A. *Garden Doctor,* Doubleday, Sydney, 1981.

SCHAUENBERG P. *The Bulb Book,* Warne, London, 1965.

THE HORTICULTURAL SOCIETY OF CANBERRA INC. *The Canberra Gardener,* 1976.

THE NORTH AMERICAN LILY SOCIETY. *Let's Grow Lilies!* 1973.

THE DAHLIA SOCIETY OF S.A. INC. *Annual Publication,* 1972.

THE SOUTH AUSTRALIAN GLADIOLUS SOCIETY INC. *The Australian Gladiolus,* 1977.

WITHERS R.M. *Liliums in Australia,* Australian Lilium Society, Melbourne, 1967.

YATES GARDEN GUIDE Collins, Sydney, 1975.

YOUR GARDEN Monthly gardening magazine. Melbourne.

Index *(General classification section)*